The Promethean Illusion

The Promethean Illusion

The Western Belief in Human Mastery of Nature

BOB TOSTEVIN

McFarland & Company, Inc., Publishers
Jefferson, North Carolina, and London

Library of Congress Cataloguing-in-Publication Data

Tostevin, Bob, 1944–
The Promethean illusion : the Western belief in human mastery of nature / Bob Tostevin.
p. cm.
Includes bibliographical references and index.

ISBN 978-0-7864-6063-2
softcover : 50# alkaline paper ♾

1. Science — Miscellanea. I. Title.
Q173.T67 2010
500 — dc22 2010034845

British Library cataloguing data are available

Manufactured in the United States of America

McFarland & Company, Inc., Publishers
Box 611, Jefferson, North Carolina 28640
www.mcfarlandpub.com

For my son Erik,
who told me I should write this book

Acknowledgments

Since this book springs from the doctoral dissertation I wrote at York University in Toronto many years ago, my thanks come in two bundles—one for those who oversaw the original study and another for those who recently have helped in various capacities to bring this work to press.

The first bundle: Thanks go especially to Christian Lenhardt, my dissertation supervisor and friend, whose trust allowed my thinking complete freedom to cut its own paths in directions of its own choosing. Thanks go also to Grey Hodnett, Ed Weissman, David Shugarman, and Brent Rutherford, all of York University, and to Alkis Kontos of the University of Toronto.

The second bundle: I am grateful to a number of my colleagues at the Pembroke Hill School in Kansas City, Missouri. Thanks go to Barbara Judd and Steve Salinger, who some time back read an intermediate draft of two chapters; to Lorraine Gordon, Kathy Ketchum, Elizabeth McIntyre, and Tom Medlock, who each read one or another portion of the manuscript and made helpful suggestions; and to Lauren Rosenfield, who provided some practical help, and especially Jeanette Jones, who has given general advice and technical assistance on a number of occasions. I also am grateful to the Pembroke Hill Alumni Association for the summer grant that some years ago provided me an entire summer to accomplish work bridging the dissertation and the present book.

Outside the Pembroke Hill community I owe thanks to Oliver Lenhardt in Toronto and Drew Bergerson of the University of Missouri at Kansas City, who each provided good advice relating to publication. But among the many who helped then or now, two stand alone: Stan Draenos, my lifelong comrade-in-thought, and Lisa, my dear wife and lifelong partner. Thanks go to Stan for always being there with probing insight and wise counsel. And thanks go to Lisa for her many loving gifts, not least of which is the room she has always made in our home for me and my thoughts to find solitude together.

Table of Contents

Preface

This book is the evolution of a doctoral dissertation I wrote in the mid–1970s when ecological consciousness was first crystallizing around newly revealed facts of our wreaking environmental havoc and placing ourselves in danger. That work drew material from the ecological crisis, but its focus was on our culture's belief in humankind's possible mastery of nature. What struck me at the time and has remained so clear to me ever since is the persistence of this belief in the face of continuing — and worsening —crisis. Indeed, programmatic proposals to master nature both then and now are entertained as solutions to ecological crisis. And then there is the more prosaic but no less telling idea that, given the right kind of institutional support, science and technology can handle nature in virtually any circumstances. My own thinking at the time stood completely at odds with such understanding. It still does.

The emergent ecological crisis opened my eyes to the essentially elusive character of nature's causalities in the life-world. So many of those causalities slip through our manipulative hands and operate unseen right before our watchful eyes. Common sense accompanied this awareness and immediately told me that nature itself ultimately eludes our grasp. Common sense told me so, with a major assist from the political thinker Hannah Arendt and the environmentalist Rachel Carson. Arendt and Carson certainly make strange intellectual companions: the one, a German political theorist whose philosophical tutelage came under Martin Heidegger and Karl Jaspers; the other, a New England naturalist given to the smells, sounds, and feel of the wilderness. Yet each from her own perspective threw light upon what has remained for me the plain and simple truth of the crisis of the natural life-world. Arendt observed from her theoretical standpoint that we are acting into nature in unprecedented ways and setting in motion natural processes whose outcomes we can neither foresee nor control. Carson revealed in her empirical studies how our multifarious intrusions into nature are corrupting the life-world and creating a massive global experiment over which we have no control. And so, via a dissertation far narrower in scope and much less historical, we come to the present book.

This work is not an investigation of the crisis of the natural life-world. Over and against the modern West's belief in humankind's domination of nature, it is the reality of nature's ultimate independence from human will that interests us in these pages. More precisely, we are concerned with the jarring collision of two realities, those being our continuing belief that nature is subject to our willful control and nature's manifest refusal to abide by this belief. Three matters needing investigation arise here. First, how is it that we in the modern West have assumed the posture of the master towards the slave in our relationship to nature? What gave rise to the belief that we can establish our dominion over this planet Earth? Second, of what does the illusion of mastery consist? What does conquest of nature mean in our most fundamental, instrumental encounter with nature, and how do we come up short? And third, why does any of this matter? How does the illusion of human dominion over this Earth blind us to what we need to see if we are to act sensibly? These are the essential questions driving this study and organizing its material.

To my knowledge there has been no systematic study attacking belief in mastery. Those who have been critical of humankind's domination of nature from the standpoint of social theory or philosophical critique generally have left the possibility of instrumental control intact. Ecologists, on the other hand, often will advance arguments about the impossibility of environmental control, but nothing in the critical ecological literature has systematically undertaken to either subvert belief in mastery or pursue the dangers the master's attitude poses for our continuing interactions within the natural life-world. This book proposes to do both. That is the argumentative cast of what follows. But this study also is an interpretation of our modern instrumental relationship to nature, analyzed through science and told as a story connecting the nature and development of modern science with Western culture. While this study is no history of modern science, it does place the instrumental character and meaning of science within a historical-cultural perspective that focuses its material on the theme of mastery. We want to understand how it is that the modern West can be so hugely successful and yet so limited, so knowing and yet so blind, in its instrumental relationship to nature.

Environmentalists and ecologists critical of the modern world's instrumental relationship to nature often fault the machine metaphor, or, more broadly, the mechanistic thinking arising with modern science. Mechanistic thinking offends nature in thought by denying it the reality of its life-world, while it abuses that life-world in practice by disregarding its intricate balance and manhandling its phenomena. The argument in these pages sympathizes with this critical concern but stays distant from it. This book is not interested in nature's well-being simply because nature transcends all such human concerns in the same way it ultimately transcends the reach of our instrumental power to command. Nature is entirely indifferent to our offensive treatment: given its scale, it is above all that; given its duration, this too will pass. Far from being any slave, nature receives our instrumental intrusions in its own time by its own measure and reacts on its own terms. The egregious mistake in our understanding and attitude towards nature is the belief that we can, in fact, command nature; that we either are or can become masters of this planet Earth. Truth be known, far from being obedient to our will, nature in many ways will have its own way with us. This book is also about this truth.

Nature's having its own way with us, at least in some ways and to some extent, is exactly what needs to be understood if we are to become truly sensible in our instrumental relationship to nature. Our illusion of mastery is dangerous because it blinds us to the nature of our real ignorance in so many cases where we act without any possible knowledge of the consequences of what we are doing, either to ourselves or to the life-world upon which we are dependent for our own well-being. In these cases nature itself will tell us what we need to know, and it will do so according to its own elusive causalities and in its own time. Arendt and Carson articulated this point so many years ago, and yet, by and large, we ultimately remain immune to its sensibility. Following their lead, this book will spotlight nature's independence from scientific command, but only after investigating how it is that the idea of mastering nature took hold of Western culture, first as legitimate belief and later as illusion.

Introduction

Western culture has never evidenced much comfort in its thinking about nature or our relationship to it. Indeed, the West's understanding of civilization betrays a decided discomfort with our relationship to nature, to say the least. The word *civilization* does not found its meaning on natural soil, but on artificial ground belonging to the city, to *civitas*, the creation of human artifice. Civil life stands beyond what is natural just as the walls surrounding cities before modern time maintained a protective distance from nature and its threats. Of course, there have been significant naysayers to such a cultural posture. One might think of Rousseau and his call for us to return to the nature within us, or, more recently, of Freud and his concern for the price the nature within us exacts for our repression of its insistent presence. Western culture itself avowedly embraced nature in the art and literature of the first half of the nineteenth century when Romanticism shaped sensibility through the umbilical cord connecting life and death in human experience to the beautiful and sublime in nature. Yet however significant such thinkers and cultural movements may have been, Western culture by and large has never been particularly hospitable to the things of nature. This discomfort with nature, in fact, has become downright adversarial in modern time, as we have taken it upon ourselves to subdue and forcefully dominate all that is natural beyond ourselves. As modern cities have expanded beyond those old walls, rendering them demarcations of *Die Altstadt* for tourists and visitors, so in modern time our emboldened stance towards the natural world "out there" has arrogantly taken upon itself the claim of conqueror. We moderns in the West generally believe that we can command nature, that we are masters of this planet Earth.

Mastery of nature is a modern, Western belief. There is little question regarding its Western character. Siddhartha Gautama surely subscribed to no such belief as he received enlightenment beneath a bodhi tree. The Buddha's path to nirvana was mastery over the self, not nature. Nor did Lao-tse have any such presumption as he prescribed for each of us the wisdom of "going with the flow," so to speak. For him, as for his societally oriented counterpart Confucius, harmony was the call of life's meaning. More broadly, neither

African cultures nor Mesoamerican civilizations could possibly have even come close to imagining domination of nature in their totemic, propitiatory religions. Humankind's dominion over nature in its origin and character is a distinctively Western belief, pure and simple. Granted that, there remains, however, the matter of its modernity. And this matter is by no means so plain and clear on the face of it.

Contrary to the position taken here, there are generally two lines of argument against the modernity of the belief in our domination of nature. Not surprisingly, given that we are dealing with Western culture, one argument looks back to Athens while the other looks to Jerusalem. The first perspective sees the idea of conquest originating in Greek philosophy and myth. Belief in human domination of nature is discernible in the Greek presumption that thinking can philosophically capture the nature of reality within a conceptual net. Entirely disparate though they are, both Heidegger's thinking and the critical theory of the founding generation of the Frankfurt School shared this insight into Greek philosophy.[1] As to mythology, Prometheus and Daedalus both point towards the idea of conquering nature in their claiming for man matters or powers belonging to the gods, and in each case an immense price is paid for such hubris. Hubris generally calls forth nemesis in Greek understanding, and nemesis suggests the very limit of Greek presumption. Puny man must not overstep the bounds established by the gods. The very idea that men might presume the powers of the gods is punishable, and pride is the very source of Greek tragedy.

Pride also goeth before the fall as we turn from Athens to Jerusalem, though here the Fall wants capitalization. Jerusalem gives us Genesis, wherein God grants to his greatest creation dominion over the fish and fowl of the Earth. The ancient Hebraic rendering of "dominion" appears open to question, however, as some suggest that the word originally had nothing to do with domination, but meant stewardship.[2] In other words, Genesis textually recalls God's charging humankind with caring for the land and animals He was placing in their hands. In any case, it is Christianity, not Judaism, that makes the telling point for those who argue from religious tradition against the notion that mastery of nature is a modern belief. Here too scriptural text is important because God creates Man in His image, which lifts humankind above everything else in Creation. Humankind thereby gains power over the lesser material of this Earth, and even allowing for stewardship etymologically, this argument can cite the Christian humanism of Ficino and Pico, early Renaissance neo-Platonists for whom the idea of dominion clearly contained the god-like presumption of the master.[3] Then there is an altogether different argument coming out of Jerusalem, one that primarily concerns history rather than the inverse anthropomorphism of God's creating Man in His image. In this view, the moderns' sense of history as a process of continuing material betterment recapitulates early (Judeo-)Christian eschatology. This argument sees a cultural linkage between Christianity's supposed "linear" view of time and the moderns' belief in progress. Standing outside of the cyclical view of time characteristic of Greek thought, early Christianity introduced history as a process in time. The problem here, however, is that the early Christians had no narrative temporally connecting past, present, and future as an ongoing process of related events.[4] Events, not process, counted; not history as such, but specific events like the Fall, Resurrection, and the Day of Judgment mattered, nothing else. But whichever argumentative path one chooses to follow out of Jerusalem, one ultimately arrives at the birth of modern science and the modern Prometheus—who is Christian, of course.

These perspectives looking towards Athens and Jerusalem do have value in the history of ideas, which is essentially a genealogical enquiry. Yet while the history of ideas may invite interest, from our perspective in this study such interest is introductory and nothing more. Mastery becomes important outside of genealogy when it takes hold culturally and roots itself among the people, or at least among large portions of those who are culturally literate. As distinct from the *idea* of domination, *belief* in domination became part of Western culture at roughly the time when the modern expansion of cities left the cities' walls behind. In point of fact, cultural belief that human beings might subject this Earth to human control for human benefit—what we call the *project of mastery*—arrives only with the rise of modernity. Nor should this fact really surprise anyone since, when all is said and done, it is actually quite difficult to imagine any premodern people anywhere assuming the posture and prerogatives of a master towards a slave while looking at nature. Such an attitude crystallized out of the scientific revolution of the sixteenth and seventeenth centuries and assumed wide cultural publicity first in the eighteenth-century Enlightenment and then in the industrial revolution that followed. In short, distinct from the idea of mastery, belief in mastery was born of modern science. It is this belief in the conquest of nature and the attitude towards nature accompanying it that, along with modern science, largely define the focus of the present work.

Stating that belief in mastery crystallized out of the scientific revolution of early modern time means more particularly that it arose upon the experimental foundation of modern science. As Francis Bacon declared at the turn of the seventeenth century in one of the world's most quotable proclamations, "Human knowledge and human power meet in one; for where the cause is not known the effect cannot be produced."[5] Resting upon experimental science, control over nature's causality is the key to modern humanity's domination of nature. Science teachers all over the world knowingly or unknowingly demonstrate such mastery in the laboratory every time they successfully command the effect of an experiment by knowing and manipulating the causality required. It was Bacon's profound insight to realize this crucial point, which makes mastery in some sense "Baconian." It fell, nevertheless, to Isaac Newton — or, better, to Newtonian science — to turn Baconian insight into instrumental fact and cultural belief. Newton's work culminated the scientific revolution of the sixteenth and seventeenth centuries, and the science arising in his name governed understanding for the following two centuries. His work delivered to the world on purely secular terms what appeared to be certain truth, and along with this truth came a practical foundation of scientific laws realizing domination of nature upon specific mathematical-experimental conditions. This dual delivery of secular revelation and mastery of nature underlay the eighteenth-century Enlightenment, and the ongoing belief that modern science authorizes human dominion over this Earth is the legacy of Newtonian science.

Newtonian science authorized modern understanding of nature outside religious traditions and sacred texts and, at the same time, transformed mastery of nature from an idea detectable in the Western intellectual heritage into a cardinal tenet of modern belief. The doctrine of progress so telling of modern belief in general entails the project of mastery; it expresses the notion that the continuing development of modern society progressively subjugates nature in expanding service to human need and public benefit. Well before the completion of the scientific revolution Bacon himself glimpsed this project and gave it fictional form in his *New Atlantis*, the first decidedly modern utopia. In this utopian fragment experimental science harnesses nature by means of "the knowledge of causes" and employs nature's own causal laws in "the enlarging of the bounds of human empire, to the effecting of all things possible."[6] Once Newton brought the scientific revolution to full fruition, Bacon's earlier vision seemed more prophetic than utopian as moderns began looking at nature through the Newtonian laws. The causality of all nature in principle lay vulnerable to experimental grasp as things appeared in the light of Newtonian authority. And beyond the testimony of science itself, as if scientific principle called upon everyday fact for confirmation, late in the eighteenth century the industrial revolution began actualizing the project of mastery in tangible ways that everyone could see and increasing numbers came to enjoy. The industrial revolution picked up steam in the late eighteenth century and advanced pell mell down Tennyson's rails of change throughout the nineteenth, materializing the project of mastery in the technological dynamism and urbanization of industrial society. With this urbanization, in turn and hardly surprisingly, neither the builders nor the inhabitants of the new cities and emerging conurbations evidenced any interest whatsoever in building new surrounding walls. They simply bulldozed nature and pushed it out of sight beneath the pavement of progress. Indeed, it did not take long for nature to become for many little more than a piece of land preserved within city parks; or, perhaps more tellingly, a place needing protection from the forces of progress so that city-dwellers might vacation there.

It is against this large backdrop that we must understand the sustainability of belief in mastery. Modern science has shaped the material structure and spirit of the entire modern age, and today that spirit finds sustenance wherever there are new frontiers to conquer: in genetic research, nanotechnology, space exploration, artificial intelligence, biomedicine, and the like. So, too, just as science has done so much to shape our material civilization, the very world we have built stands for our commanding presence on this Earth. Be it in research laboratories, in technologies of production and information processing, or in the steel and glass of our urban environs, human artifice seems everywhere triumphant. Triumphant indeed, and standing there behind this construction, like a master architect, is the mathematical-experimental science that combined knowledge and power in good Baconian fashion to materialize our modern world. Amidst the very astonishing success of such construction, there seems little room to doubt human dominion over the Earth. And for all too many people today, there is comfort in there being little room.

Sometimes contemporary belief in mastery boldly proclaims itself, as we hear in reports of our verging on possession of all kinds of genetic capabilities regarding the elimination of disease, the enhancement of the human being, the indefinite postponement of aging and death, and with all of this, the design and control of our species' evolution. Sometimes belief in mastery insinuates itself more subtly, as in the not narrowly held assumption that, given whatever is needed in the way of political and economic support and sufficient time, science "can find the answer" to virtually any problem nature might pose. Nor is such confidence without the professional support of many experts themselves. If we are to believe Richard Lewontin, a population geneticist who is never anything but forthright, "Natural scientists, in their overweening pride, have come to believe that everything about the material world is knowable and that eventually everything we want to know will be known."[7] Of course feeding such belief, be it held by scientific experts or by ordinary citizens, are many of the futurists and science writers who are on the reporting beat for this, that, or the other journal, magazine, or video outlet, just looking to publicize the latest and greatest, or most promising "advance" to dazzle us. Constantly we learn of new wizardries on virtually all research fronts. And to what effect are such wizardries reported, if not to reinforce belief in mastery via continuing evidence of an amazing progress having no limits?

Beneath the triumph of modern artifice and the continuing wizardries of instrumental knowledge lies another factor sustaining belief in our domination of nature, a grassroots element of belief, as it were. At this grassroots level knowledge and power meet most fundamentally in the manipulative conditions of scientific practice where belief in mastery unites with the scientific attitude experimentally taken towards the things of nature. It is a commanding attitude born of the very success of modern science. Once experimentation evidenced nature's submission to the rule of the master, it then became quite easy to suppress and forget nature's essential independence from human will by simply taking its submission for granted. In other words, it became easy to presume that nature has nothing to say about whether or not its causal secrets are experimentally discoverable and instrumentally manipulable. This presumption takes as its model the relationship between master and slave belonging to the modern institution of human slavery. Within that modern relationship the master himself set the rules to which the slave obediently had to submit. The slave had no voice in any law-making apart from the benevolent consideration a particular master might show his slave. In any case, the master determined what was allowed; he brooked no contradiction of his will. Modeling our scientific relationship to nature on this basis—assuming the attitude of such mastery in the case of nature's causality—however, commits understanding of our relationship to nature to an egregious fallacy. For in this relationship it is not we who write the laws, but nature itself. As Bacon himself wrote, in the same short aphorism where he announced the meeting of knowledge and power, "Nature to be commanded must be obeyed...."[8]

Obedience to nature in principle would pose no problem for human dominion if nature's causality were uniformly exercised according to Newtonian understanding. As long as that understanding might prevail, there was no reason to assume anything but unobstructed progress in our continuing exercise of experimental command over the natural world; and since scientific understanding meant truth as well as mastery, consigning to oblivion Bacon's point of our having to obey nature seemed non-problematic. But scientific progress would prove thoroughly ironic as the development of Newtonian physics in the course of over two centuries undermined the claims to both truth and mastery. Those claims dissolved in the early decades of the twentieth century when physical science advanced beyond Newtonian understanding, only to find nature unwilling to yield all of its secrets or be experimentally commanded as if it were a human slave. Later in that century nature voiced even larger objections to the presumption of mastery outside of physics, where biological and ecological phenomena possess causalities not always reducible to neat, manipulable variables. The thrust of those developments, first within physical science and then in biology and ecology, evidences an inverse corollary to Bacon's point of obedience: to wit, *and where nature will not allow its causality to be controlled, we must concede nature's independence from experimental command.* We phrase this inverse corollary to Bacon's dictum so as to capture what nature has been telling science for some time now, namely that its causality is not uniform and that it will not necessarily yield its operations to experimental control. Yet we in the West, or at least all too many of us, have not paid much attention: the master's ears have been deaf to nature's voice.

What follows in the early chapters of this study is the story of how Western culture came to believe that science authorizes human dominion over this Earth. Part cultural interpretation of modern science, part theoretical analysis of the nature of mastery, our narrative itself in no way approaches a history of physical science, not that there is any qualification here to presume the writing of such a history in any case. Having completed that story, this study in its later chapters then turns to specific biological and ecological case studies that strikingly serve to illuminate the illusory nature of the continuing belief in mastery. No more than our narrative presumes to be a history of science does our treatment of these case studies pretend to be anything nearing an analysis of what lies behind the current crisis of the natural life-world. Our interest in those empirical studies is simply to demonstrate the inverse corollary to Bacon's dictum and, in so doing, to tackle head-on the two major projects of mastery programmatically being proposed today, they being genetic engineering and planet management. In the end this study's treatment of science and technology will be completely one-sided, its focus being restricted entirely to the limitations of instrumental knowledge. Nothing in this work, however, abuses science or calls for Luddism in any way, shape, or form. Let there be no misunderstanding: the aim here is only to undermine the belief that science authorizes our domination of nature and to expose that belief as illusion.

Anticipating the substance of this study as best we can at this point, one fundamental matter may be elliptically suggestive. We begin our story of modern science with the "quest for certainty" that philosophy historian Richard Popkin deems the central and defining preoccupation of early modern European thought.[9] As religion lost authority to govern understanding in early modern time, men sought new ground on which to certify the truth of their knowledge claims. Our study roots its cultural interpretation of science as well as its analysis of mastery in that quest for certainty and sees that quest's intellectual and cultural fulfillment in the secular revelation and experimental control of nature arriving with Newtonian science. Fast-forwarding now through the development and dissolution of Newtonian science to the mid-twentieth century, our culture once more returns to uncertainty. But this return is by no means circular. Whereas modern science at its birth dismissed uncertainty in its claims to truth and mastery, our contemporary uncertainty already has science in it; and, moreover, this very uncertainty manifests our inability to command nature. Our study will closely examine the connections between certain knowledge and mastery on the one hand, and between uncertain knowledge and the absence of command on the other. In the end, short of whatever particular insight this study may offer into our current interactions with the life-world, the present work claims to be a brief but coherent cultural-historical account and analysis of the modern West's instrumental relationship to nature.

1

Mastery in the Making

The life and thought of sixteenth- and seventeenth-century Europeans lay on the hinge of time, for theirs was an experience lived in twilight, in a time between ages. For the late medieval world passing into darkness, it was dusk; for the modern age just coming to light, it was dawn. In such half-light there was little that people could see clearly in common. Anarchy threatened in the realm of belief precisely because no common authority obtained any longer.[1] No longer did the Church of Rome govern the province of European understanding; traditional authority had been eclipsed. The Protestant Reformation had indeed fragmented the very meaning of Christian belief, rendering questions of truth and claims to knowledge sectarian. And if the Reformation needed allies to complete the disintegration of Western understanding, help certainly was forthcoming in the work of Copernicans, whose speculations and discoveries increasingly threw the long-standing Christian-Aristotelian cosmology into doubt; and in the work of global explorers, whose continuing reports of theretofore unknown life forms and newly found cultures overturned many older assumptions about the world and fostered a form of cultural relativism among the more sensitive souls. Amidst such earthshaking speculations and unsettling novelties, there arose deep and widespread uncertainty about what we actually do know about what is real and what is not. Such uncertainty, at once psychological and cultural, was itself that twilight world in which men's understanding lacked common bearings.

Working in that cultural milieu and witnessing the disintegration of old beliefs and traditional modes of comprehension were legions of thinkers who found in such disarray the very reason to think. Numerous thinkers saw in the disagreements dividing people the impossibility of ever establishing any truth, and skepticism enjoyed a new heyday. Many others, thinking differently and wanting to see their way clear of such cultural confusion, sought in various ways and in numerous intellectual disciplines to establish new foundations for human understanding. The key issue in such pioneering efforts was the question of agreement, or common assent: how could people come to some binding agreement on knowledge claims about what is real or what is true?

Without any religious accord, on what ground might early moderns commonly stand in their need to speak knowingly of the world around them? Pursuing answers to these questions constitutes the "quest for certainty" that historian of philosophy Richard Popkin deems the chief animating force behind early modern European thought in general.[2] That quest for certainty provides us the compass to be followed in this first chapter, as it will lead directly to the mathematical objectification of reality and to the programmatic manipulation of nature's things. The wonderfully fruitful marriage of these two achievements brought modern science into the world, and with it came authorization of the belief that humankind might exercise dominion over this Earth.

Many mathematical minds and manipulative hands went into this wedding during the time between the middle of the sixteenth century, when Copernicus proffered his heliostatic picture of the universe, and the end of the seventeenth century, when Isaac Newton delivered modern science to the Western world. Most historians of science refer to that gestation period of science as the "scientific revolution" of early modern time, and we will give some brief attention to the scientific revolution in the following chapter.[3] Our interest in this first chapter, however, lies not with the actual emergence of modern science. Our present focus, rather, will be on the two parties wed in that science: the mathematical objectification of reality, which is most clearly understandable in the work of René Descartes, who lived through the first half of the seventeenth century, and the experimental manipulation of nature's things, which is best grasped in the program for learning that Francis Bacon articulated in his *New Organon* early in that same century. We want to see how Descartes and Bacon each bound the search for truth to the interest in mastering nature.[4] Whether or not this common enterprise predisposed the objectification of reality and the manipulation of nature's things to marry does not concern us as much as realizing that when modern science came into the world in Newton's work, it was no accident at all that its birth certificate combined a mathematical claim to certain knowledge with an experimental claim to controlling nature.

Doubt and Uncertainty in Early Modern Time

Investigating Descartes' reflections and Bacon's program for learning first requires our locating their enterprises within the larger context of early modern experience. We need to root their quests for new understanding in the uncertainty surrounding those quests and giving rise to them. And at the beginning of the seventeenth century John Donne versified that uncertainty in oft-quoted lines that spoke directly of the impact Copernicus' ideas had been having on more than just a few:

> And new philosophy calls all in doubt,
> The element of fire is quite put out;
> The sun is lost, and the earth, and no man's wit
> Can well direct him where to look for it.
> And freely men confess that this world's spent,
> When in the planets, and in the firmament
> They see so many new; then see that this
> Is crumbled out again to his atomies.
> 'Tis all in pieces, all coherence gone....[5]

Donne's lamentation contextualizes the loss of coherence historically, within the experience of early modern time, but for others like Blaise Pascal, writing a few years later at mid-century, uncertainty revealed the tragic nature of the human condition irrespective of time or place. Whereas Donne's lament was that of an Anglican clergyman, Pascal's reading of uncertainty sprang from his Jansenism and expressed more than a little Augustinian anguish: "Such is our true state.... We are floating in a medium of vast extent, always drifting uncertainly...."[6] Doubt, uncertainty, and incoherence all followed from the Fall, and, for Pascal, the experience he shared with his contemporaries simply made starkly manifest the inescapable state in which human beings—sinners all—are caught and must suffer.

> We desire truth and find in ourselves nothing but uncertainty.
> We seek happiness and find only wretchedness and death.
> We are incapable of not desiring truth and happiness and incapable of either certainty or happiness.
> We have been left with this desire as much as a punishment as to make us feel how far we have fallen.[7]

Donne's disorientation and Pascal's anguish by no means exhaust the early moderns' interpretation of the meaning or nature of uncertainty. Looking further, we find another Catholic thinker, Michel de Montaigne, reflecting on the nature of uncertainty. Unlike Pascal's despondent bent, however, Montaigne's late-sixteenth-century reflections actually follow the flow of the time without any religious regret for the human condition.[8]

Far from suffering Augustinian distress, Montaigne's intellectual sensibility fully resonated with the worldly experience of his time, addressing and prismatically reflecting the plurality of Christianities, the conflict between Copernican and Ptolemaic astronomical perspectives, and the cultural relativity suggested to him by a European anthropology that was continually expanding with the voyages of exploration and discovery. As Montaigne himself understood things, his own religious allegiance reflected the adventitious intersection of biography and history. His believing in the Church of Rome and the traditional truths it upheld constituted a colossally profound accident of birth; had he been born at a different time or elsewhere, his religious convictions would have been just as unquestionably staunch as convictions, yet, at the same time, altogether different.[9] Cultural relativity, however, supplements rather than frames Montaigne's perspective. Culture may diversify beliefs customarily, but for Montaigne uncertainty springs inevitably from the nature of human experience. "How our judgments vary! How frequently we change our ideas! What I hold and believe today, I hold and believe in the totality of my beliefs.... But," he goes on, "not once, not a hundred times, not a thousand times, but every day—have I not embraced something else with the same resources and under the same circumstances, only to be convinced later that it was wrong?"[10] Articulated in experiential plain-speak here, Montaigne's skepticism appears psychological. But Montaigne clothed such naked sensibility in the philosophical arguments of the ancients, giving philosophical currency to the pyrrhonian skepticism of Sextus Empiricus, a third-century Greek whose texts had been recovered in Montaigne's own time, thirteen centuries later.

Montaigne does not simply proclaim that our judgments may vary from one day to the next. Following the pyrrhonism of the Greek skeptics, he proceeds on various philosophical planes to argue the inherent defectiveness of our judgments and beliefs.[11] Our direct or immediate experience of things offers our judgment no firm footing. "We register the appearance of objects; to judge them we need an instrument of judgment; to test the veracity of that instrument we need practical proof; to test that proof we need an instrument. We are going in circles."[12] Nor can reason save the senses from such uncertainty. In fact, reason suffers a similar fate. Reason is insufficient to ground truth or establish certain knowledge because it demands a reason for every premise; each reason, once given, in turn becomes yet another premise standing in need of a further reason, such that we fall into an infinite regression. In general, as Montaigne concludes, "There is no permanent existence either in our being or in that of objects ... [and] nothing certain can be established about one or the other, since both judged and judging are ever shifting and changing."[13]

Just as Pascal's aphoristic style befit his own disbelief in the provability of any system of truth, so Montaigne's creation of the essay as a literary form bespoke his skepticism: doubters do not build large intellectual structures. Montaigne articulated, nevertheless, a ratiocinative set of arguments that established early modern skepticism on firm philosophical footing and made it widely employable in time of doubt. And skepticism was in great demand because, ironically, it was useful to any and all partisans of the theological conflicts that seemed to leave doubters out of the picture. Pyrrhonian arguments against the other side's knowledge claims became weapons in the rampant religious warfare being fought on theological-philosophical fronts as well as on bloody battlefields from the early sixteenth century right up through the first half of the seventeenth.[14] While on those bloody battlefields many died with religious certainty in their minds, in these theological trenches bad faith obtained across the entire front: how could any party be certain of its own truth while wielding the weapons of skepticism against the enemy? Popkin writes of the terms of the basic issue as follows: "The attack begins with the problem of the criterion raised by the Reformation; how do we tell what is the rule of faith, the standard by which true faith can be distinguished from false faith?"; or, more specifically, "Who is going to tell what Scripture says? It is here that a dispute exists, not just between Catholics and Reformers, but between Luther, Zwingli, and Calvin as well."[15]

Radiating outward from the religious earthquake caused by Luther's successful defiance of Rome, divisions among European Christians made partisan any and all appeals to scriptural truth. Luther's belief that individual conscience or inner light should guide each believer to such truth offered no bulwark at all against differences of meaning or conflict among scriptural interpretations. It was one thing to claim as reformers did that truth is to be found in the Bible, but quite another to find such truth outside of institutionally based authority like that provided by Rome. Protestants, in fact, bore witness against themselves in their own sectarian disagreements, thus confirming in appearance at least the Catholics' charge that the reformers' reliance on individual faith invited anarchy in Christian belief. Skepticism was a two-edged sword, however, and Catholicism fell victim to the same kinds of arguments employed by Rome.

Hardly safe from attack was the criterion of an institutional tradition on which the Catholic cause rested its own claim to interpret scripture authoritatively. Reformers aimed their argumentative logic towards the infinite circularity of the Catholic position, demanding to know by what fixed and unchallengeable criterion the criterion of an institutional tradition was itself to be authorized. What hallowed the foundation of Roman authority apart from interpretation? Rome's institutional position may have weathered this war of words, but the war itself nevertheless directly exposed the fractured foundation of Roman authority. In the end, as throughout the time of religious warfare, the question of how truth could be religiously grounded remained unresolved. Popkin sums up the situation as follows: "The intellectual core of this battle of the Reformation lay in the search for justification of infallible truth in religion by some sort of self-validating or self-evident criterion. Each side was able to show that the other had no 'rule of faith' that could guarantee its religious principles with absolute certainty."[16]

What did emerge from these conflicts was the steadfast character of faith itself. Fideism—faith without proof—remained the essential reality of early modern Christian skepticism. Pascal took issue with all of those who, be they skeptics or dogmatists, went about their faith by argumentative means, as if God's truth were to be argued out one way or another and a person's faith were somehow consequential to such argumentation. For Pascal, faith belongs to the human heart, and as he quite famously declared, "[T]he heart has its reasons of which reason knows nothing...."[17] The fideism of the early moderns did not, however, follow from any pyrrhonian reasoning, and their faith was anything but heartless. Faith without proof was in point of fact the very well-spring of their intellectual sensibility. The nineteenth-century philosopher Hegel understood this point clearly when, centuries later, he perspicaciously distinguished between ancient skepticism and the modern version. Whereas ancient skepticism consisted in thorough-going, all-encompassing doubt and constituted a "thought out annihilation of everything which is held to be true and existent...," modern skepticism entailed "the subjectivity and vanity of consciousness.... For it goes no further than saying, 'This is held by me to be true, my feeling, my heart is ultimate to me.' But here certainty is alone in question, and not truth...."[18]

Hegel's philosophical vision here helps us to illuminate the large cultural landscape of early modern time. Ancient doubters had swept away ontology, and when Being disappears, it takes everything with it. Early modern doubters, on the other hand, left Being intact: Being meant God. The existence of the Christian deity met no challenge in the deep differences of Christian belief marking the theological conflicts of the sixteenth and seventeenth centuries. God and His Creation were entirely real: of that there could be no doubt! What had become problematic was the ground on which people stood when claiming to speak knowingly about God or the things He had created. The disintegration of common religious ground in the Protestant Reformation meant that scripture itself had lost its authority to bind people in common understanding. Once Rome lost its institutional hold on people's allegiance, scripture and all canonical texts became widely subject to what individuals or sectarian groups interpreted them to mean. Such subjectivity is precisely what Hegel refers to as the vanity of consciousness: "'This is held by me to be true.'" And as he further observes, what is at issue here is not the truth of Being but the certainty of knowledge: amidst all of these differences of meaning and disagreements of understanding, how can one authoritatively substantiate knowledge claims? In asking this question the early moderns gave birth to modern epistemology.

Matters of epistemology, or the issue of how knowledge claims are to be grounded, had long lain latent or implicit in the metaphysics of the religious and philosophical traditions, but those traditions had hegemonically suppressed such matters within the textually categorical confines of Being and Truth. The rift in the foundation of common religious authority splintered truth and lifted epistemological questions out of such age-old categorical incarceration. Early modern Europeans were still Christian, and deeply so, but the dissolution of common scriptural authority threw men's thinking back upon itself, giving rise to epistemology, or the issue of how we can know what we claim to know, in the recoil. Religious stalwarts might continue clinging fast to particular scriptural grounds of understanding out of strict institutional allegiance, sheer unyielding will, or just plain psychological comfort in the midst of profound and often bloody disagreement. In any case, the theological-philosophical front of the religious wars had itself disclosed the inadequacy of scripture to bind people together authoritatively.

Pyrrhonism provided no viable alternative to this situation. Could skepticism have prevailed as the way of the world, moderns would have lived their lives like Plato's cave-dwellers, subject to conflicting opinions without the possibility of any truth to bind them in their shadow world. But that was not to be. There were some cave-dwellers, with initial doubters among them, who turned their backs on the skeptical premise that uncertainty is the inescapable condition of all knowing and understanding. Like Plato's philosophers, they knew that appearances could not be trusted, and, without succumbing to skepticism, they sought some ground for ascertaining what lay behind appearances. Unlike Plato's philosophers, however, they sought no *eidos*, no transcendent realm above and beyond the world we commonly inhabit. Whatever lay behind appearances was, in fact, within the cave in some sense. These cave-dwellers remained where they were in their respect of minding how everyday appearances might be conditioned or made knowable in some sense. And from within their ranks came those who directly or indirectly contributed to the founding of modern science.

These seekers of certainty unsurprisingly held hands with their God in their work, yet there was nothing religious in the materials they used to fashion their epistemic foundations. While wanting to glorify God in what they did, they themselves glorified the very enterprise of establishing a way of authorizing knowledge claims without any appeal to tradition, religious or other. Historians of science may trace genealogical connections of all sorts between what went into the making of the scientific revolution and what previous thinkers had thought and earlier doers had done, but the critical fact is that scientific authority legitimized itself and compelled common consent on entirely new epistemological ground having virtually no regard for the past.[19] In the next chapter we will closely examine the nature of such authority, but in what directly follows here, we want to investigate the epistemological ground of modern science in terms of the constituent elements originally defining it: the mathematical objectification of reality and the experimental handling of things. We need to look, more particularly, at the work of Descartes and Bacon, whose thinking illuminates better than any other how the quest for certain knowledge and the mastery of nature converged in the attempt to establish understanding anew.

Attempting to Govern the Realm of Appearance

Whereas Pascal and Montaigne turned their backs on any prospect of erecting new systems of understanding, others sought to lay new foundations upon which structures of knowledge might be systematically built. Many of the most prominent intellects of the seventeenth century were stonemasons working on one kind of foundation or another. Hobbes, Descartes, Locke, and Bacon, to name but some of the more influential, worked on different grounds using different materials, but of the various early modern stonemasons, it is unquestionably René Descartes and Francis Bacon who are most germane to our story of modern science. It is not at all uncommon to encounter their coupling in the traditional literature treating the intellectual origins of the early modern scientific revolution. Indeed, discussions of the rise of modern science sometimes address Newton's work as a synthesis of the contrasting epistemological perspectives to be found in Descartes and Bacon.[20]

Those familiar with detective fiction will immediately grasp the traditional approach to Descartes' and Bacon's different perspectives by considering Agatha Christie's detective character Hercule Poirot and Conan Doyle's famous Sherlock Holmes. Poirot continually must tell everyone he is Belgian despite his name and sensibility, which appear French; but the reality is, Christie has intentionally created a character in the mold of French rationalism. After Poirot has picked up whatever clues he deems relevant, he likes to retire to an armchair or maybe to a table, where he may steady his nerves by building a house of cards: there and then, *voilá*! He uses what he calls his "little grey cells" to order reality and penetrate the truth lying behind the appearance of things. Conan Doyle's famous detective Sherlock Holmes, on the other hand, proceeds from start to finish looking for visible clues that ultimately will lead him to the truth. Holmes' close observation of things is symbolized by the magnifying glass, and he will elicit from a person's garment, perhaps a hat, a complete picture of that person's age, socio-economic background, present financial status, psychological predispositions, etc. In contrast to Poirot's ratiocinative brilliance, Holmes' genius keeps always in direct observational touch with what the world offers him. Poirot is the French rationalist *par excellence* whereas Holmes is the English empiricist pure and simple. Translating these different methods of operation into epistemological approaches gives us the standard "rationalistic" and "empirical" readings of Descartes and Bacon that not atypically have entered texts on the scientific revolution.

Descartes and Bacon each featured himself a pioneer, presented his work as a clarion call for new beginnings, and believed his work to establish for the very first time a clear and certain path to truth. Neither thinker by his own light entirely saw what modern science ultimately would require or entail in full, yet each anticipated Newton and Newtonian science in a way that illuminates the epistemological foundation of modern science. Turning to their work now, we need, however, to be clear that our own interpretive tack will differ somewhat from the traditional readings. First, we do not want to appear in league with the idea that Newton somehow married Cartesian/French rationalism and Baconian/English empiricism. Even apart from other possible criticism, this idea is simply wrongheaded in its empirical reading of Bacon, and we will critically treat this problem in later discussion of that Englishman's work. More generally, we may note at this point that we depart from the normal emphasis highlighting contrasting epistemologies. Descartes' and Bacon's *modi operandi* certainly do differ, as will be plainly evident in the discussion to follow, but our basic interest is to spotlight the animus common to their intellectual enterprises. For each thinker, the quest for certainty means controlling the realm of appearance in some way so as to insure that our knowledge claims are absolutely certifiable.

However much Descartes' and Bacon's quests for certainty may differ, they do entail the same driving force, which is nothing less than controlling how things appear; only then might one be sure of the true reality behind or within those things. Each thinker sought to establish a realm in which the very appearance of things is itself governable; to establish, that is, a domain in which human command is the *sine qua non* of appearance itself. This point needs remain a terribly abstract idea at this point. Yet we need to understand that while Descartes and Bacon surely have in mind very different realms of appearance when they speak of *how* or *where* we can truly comprehend the reality of nature's things, neither of these thinkers has in mind the literal, everyday domain of appearance—the world given to the senses in which the sun appears to move through the firmament and sticks appear to bend when thrust into water. Indeed, the concern many early modern thinkers had to establish new foundations for understanding had everything to do with their rejection of the everyday world of appearance: the everyday world where the sun, in fact, truly does appear to move across the sky, and sticks, in fact, truly do appear to bend in water. Quite simply, the senses could not be trusted. Early modern pyrrhonists drew from the ancient skeptics many arguments for distrusting the senses, but it was Nicolas Copernicus who more immediately and much more meaningfully offered early moderns fresh conjectural evidence for such distrust when he proffered them his heliostatic universe.[21]

Copernicus is a key figure in the scientific revolution. By proposing his heliostatic universe in the place of Ptolemy's geocentric version, that sunstruck Pole articulated a perspective that not only contradicted the traditional Christian-Aristotelian understanding of things, but which also undermined the empirical standpoint from which reality itself might be assessed. As Copernicus at one point saw fit to denigrate the literal, everyday appearance of things, "[T]he principal attempts by which the natural philosophers attempt to establish the immobility of the earth rest for the most part on appearances; it is particularly such arguments that collapse here, since I treat the earth's immobility as due to an appearance."[22] The "Copernican revolution" launching early moderns in the direction of science immediately contraposed reality to appearance and thus directly rendered "appearance" problematic. Descartes himself aimed to surmount uncertainty by reconstituting the realm of appearance on the ground of pure reason, where rational thought alone might realize "objectivity" by governing what was ascertainable in terms of logic and mathematics. Eschewing mathematics, Bacon, on the other hand, fashioned the realm of appearance with his hands, creating experimental contexts in which nature's causality might be discovered and controlled. Taken together, their work directly illuminates the role and nature of mastery at the very origins of modern science.

In his "First Meditation" Descartes puts the problem of appearance this way: "Everything which I have thus far accepted as entirely true (and assured) has been acquired from the senses or by means of the senses. But," he goes on, "I have learned by experience that these senses sometimes mislead me, and it is prudent never to trust wholly those things which have once deceived us."[23] Descartes' experience of uncertainty is one with the fact that he cannot assume the reality of what appears through the senses. This experience is primordial for Descartes; it underlies all that follows in his philosophical reflections. And what follows immediately upon this uncertainty is absolute senselessness: turning away from sensory experience altogether, Descartes closes his eyes and shuts himself off entirely from the common everyday world given to us through the senses. There, in the worldless darkness of that utter senselessness, Descartes radicalizes his uncertainty to embrace his own existence. If he cannot see himself or feel himself with any certainty, how can he know that even he, René Descartes, exists? His answer to this question articulates his first principle of philosophy, upon which he bases all further knowledge claims. That first principle is "*cogito ergo sum*"—"I think therefore I am." This principle is the most famous philosophical utterance of modern time, and rightly so since it laid the foundation of modern Western philosophy in human consciousness while realizing mind-body dualism as the central problem of human self-understanding in the modern West. For our purposes, however, Descartes' proposition reads better in its fuller formulation, which expresses the experiential root of his meaning in its historical context: "*Je doute, donc je pense, donc je suis*"—"I doubt, therefore I think, therefore I am." This "I" of "I am" is the Cartesian *cogito*.

The *cogito* is a thinking being that possesses no sensory apparatus of any kind; it is absolutely disembodied and consists entirely of reason's self-reflection. Descartes conceives himself to be this purely rational being in his escape from radical doubt, and his identity as a rational being is his first demonstrable certainty.[24] In point of fact, however, the demonstration does not work. Descartes' first principle is logically flawed by the circularity of his assuming in the first place (the I of "I doubt") what the predicative logic is meant ultimately to demonstrate (the I of "I am"). Yet this point bears no significance apart from logic. More telling is Descartes' essential concern for logical proof. In the worldlessness of pure thought, reason can and must establish certainty on its own terms, which in this case means logical demonstration. Descartes articulates the *cogito*'s very existence on these terms, the fallacy in his demonstration notwithstanding, and so his thinking in this way establishes a realm of appearance in the domain of pure reason, a realm which rational reflection can and must govern only on its own terms.

This cogitative ground of Cartesian philosophy may seem far removed from anything relating to modern science, but our discussion here is hardly digressive. At the heart of modern science lies the distrust of the senses we have already considered, and Descartes' reflections speak directly to the "objective" ground of scientific understanding—indeed, they articulate "objectivity" as a solution to that distrust—as they unfold from the senseless *cogito* to the material world surrounding it. And back to the material world Descartes' reflections must go, of course, if the *cogito* is to speak knowingly of the reality of those things from which it is primordially disengaged. Descartes must build a bridge back to the very world of the senses to which he initially closed his eyes and turned his back, and the constitution of this bridge needs be of the same stuff as the *cogito*. That is, Descartes must find his way back to the everyday world and all of its material contents by way of pure reason's own terms, which are meant to govern all appearances in the interest of certain knowledge. Reason's own terms are logic and mathematics, which reason produces out of itself and which reason can use as a bridge to return to the everyday world of appearances.

Appearances are to be grounded upon purely logical-mathematical terms, or what Descartes categorically calls "clear and distinct ideas."[25] Clear and distinct ideas are those in which reason can demonstrate rational certitude. Descartes' first principle at least formally exhibits such character in its propositional structure. More generally, logic entails rational certitude precisely because it is a product of reason designed to validate or certify propositional thinking. Validity means certainty in the movement of thought. Like logic, and integrally bound to it, geometry and mathematics in general also are products of reason and bear certainty in the axiomatic structure and rules of reckoning governing their formulas. Along with logic, then, geometry and mathematics fill out the meaning of clear and distinct ideas and constitute the ground upon which the *cogito* can open its eyes to the world of things materially surrounding it. The *cogito*, or the *res cogitans* (the thinking thing called man or woman), defines nature's appearances as *res extensa*, extended things having width, height, weight, volume, position and other measurable properties. Nature is thus reduced in such a way that things are knowable only insofar as they are mathematically constituted in their appearance. Hence, our knowledge of things and the domain of appearance converge on mathematical ground within the concept of magnitude.

> The principle function of human investigation consists only in reducing these proportions of things to such a form that we can clearly perceive an equality between the entity sought for and something else which is known. It is then to be observed

> that nothing can be reduced to this equality unless it can exist as more or less, and all this is included in the term "magnitude." In consequence, we recognize that ... we are, in the last analysis, concerned only with magnitude in general.[26]

Here we get our first real glimpse of modern science, or at least of one of the cornerstones of the science that emerged in the intellectual-cultural revolution that unfolded in the sixteenth and seventeenth centuries. Modern science is visible to no small extent in this mathematical reduction of all things "real" to magnitude.

From Descartes' own point of view, mathematical understanding simply lays hold of a truth, giving witness to the reality of God's Creation: "[E]verything which I conceive clearly and distinctly as occurring in them [nature's things]—that is to say everything, generally speaking, which is discussed in pure mathematics or geometry—does in truth occur in them."[27] Galileo meant essentially the same thing where he famously declared that anyone who would read the Book of the Universe properly, would have to do so in the language of mathematics.[28] But this understanding of what is immediately real in the appearance of things—this is verily the way things are!—actually assumes a reality that already has been mediated by human thought, or in Descartes' particular case, by clear and distinct ideas. Descartes admits as much at another point, where the primordial worldlessness of the *cogito* shows itself in some disdain for direct observation. Descartes writes there that "we ourselves ... compose those things which we understand, in so far as we believe that something is contained in them which is immediately perceived by our mind without being found in experience," and elsewhere that "[that] which in my nature is called perceiving ... speaking precisely, is nothing else but thinking."[29] Perceiving is thinking, and thinking is objectifying. Descartes' quest for certainty has led us to the objectification of nature.

The objectification of nature consists exactly in the mathematical grounding of nature's appearance(s) under the rule of magnitude. Under such rule, nature's things obtain objective reality—they become real—only insofar as they appear mathematically. Immediate with this, nature's things become mathematical objects in their very appearance. The broader implications of such objectification were clear at the time and found widespread articulation in the doctrine of primary and secondary qualities that Descartes, Galileo, Locke and other early moderns saw fit to promote as a kind of theoretical gloss on the encounters we have with the world.[30] According to this doctrine, most generally considered, primary qualities are the objective properties of things, as these properties are ascertainable under the rule of magnitude. What can be measurably ascertained is objective; it is real in its appearance. Secondary qualities, on the other hand, are those matters that elude such quantitative determination and thus fall short of objectivity; they are not real "out there," so to speak. Matters such as smells and tastes are inner impressions, subjective states, expressing our inner experience of things out there. Not yielding to the rule of mathematical constitution, smells and tastes are banished to the human interior; they disappear into the self where, considered as things experienced, their existence is altogether private or personal in some uncertain sense. The subjective simply does not appear. The doctrine of primary and secondary qualities thus generally sorts out what appears and what does not appear on the basis of the objectivity of things versus the subjectivity of human experience. In this doctrine one can catch foreshadowings of the nineteenth century's Romantic reaction to the imperial claims of science as well as C. P. Snow's mid–twentieth century idea of there being a divorce between "two cultures," that of the sciences and that of the humanities.

The doctrine of primary and secondary qualities helps us to distinguish the modern mathematical rendering of reality from the premodern mathematical rendering, which belonged largely though not entirely to the (neo)Platonic or Platonic-Pythagorean traditions. Plato's philosophical topography no longer holds: there is no longer above the cave of the everyday world and opinion a transcendent realm of the *eidos* where the truth of the Being of things can appear in contemplation. Mathematical reason in modern time does not prepare the philosophical soul either for mystical numerological ascent (Pythagoras) or for the ethical and aesthetic contemplative embrace of transcendent Being (Plato). Mathematical reason itself now transcends opinion and/or subjective impressions within the cave by way of objectification. Rendering appearances objective so as to provide for certain knowledge is the mark of modern scientific reason.

The thinking of many early moderns did have neoplatonic or Platonic-Pythagorean roots, and, indeed, many historians of science point to those roots in addressing the scientific revolution's mathematization of nature.[31] This understanding is hardly wrong in the case of a Kepler or a Galileo, but locating the essence of modern scientific understanding among the Greeks tends to mislead us by conceptually flattening the meaning of the mathematical understanding of nature and historically assigning to the mathematics of modern science a premodern functional role. Descartes' work, not Galileo's, is the telling case here, for the crucial point — the modern turn of mind, as it were — arises with the objectification of nature, which Descartes' work makes crystal clear. Nature's objectification comes with the knower's investigation of the grounds of knowledge, an investigation undertaken in the interest of establishing certain knowledge. Epistemology therefore mediates ontology in the modern mathematical objectification of nature. This philosophical move is of fundamental importance to modern understanding in general. In the *res cogitans* and the *res extensa* Descartes articulates the *subject* and the *object* that define the intellectual coordinates of human self-understanding and the understanding of nature in modern Western discourse. And needless to say, that discourse over these past few centuries has been anything but Platonic or Pythagorean.

Descartes' thinking helps us to situate the objectification of reality upon the experiential terms of early modern doubt and uncertainty. His thinking at the same time helps illuminate the integral connection between the objectification of reality and reason's claim over the domain of appearance. In modern science the objectification of reality is in and of itself an attempt to control the way in which nature's things are to be seen and understood. But reason's claim to govern appearance alone, purely on its own terms, is inadequate to controlling nature in fact, and the objectification of reality would prove necessary but insufficient to the founding of modern science. While Descartes saw in the *cogito* a new foundation for knowledge and spoke directly of *res cogitantes*' becoming "masters and possessors of nature" by way of the new "practical philosophy,"[32] he himself really gave relatively slight attention to either experimentation or the practical sphere of applied knowledge. Indeed, the disembodied nature of the *cogito* from the very outset made problematic any effective governance of the realm of appearance. Mathematical reason might condition the realm of appearance in the selfsame way objectification would allow or disallow the appearance of certain aspects of nature's things, but inasmuch as the disembodied *cogito* possesses no hands to hold nature to account, reason alone would not be able to govern the realm of appearance. If nature were to be not just an object, but a controllable object, reason would need the work of men's hands. In other words, the objectification of nature wrought by mathematics would need a realm of appearance fashioned by experimental design. And just as the objectification of nature found clearest articulation in the work of René Descartes, so in early modern time the experimental handling of nature found its staunchest advocate in Francis Bacon, who sought to control the realm of appearance in a manner quite different from Desacartes' approach.

Whereas the *cogito* presumed the prerogative of an absolute monarch in assuming that reason alone might legislate appearance as a law unto itself, Bacon urged democratic rule by insisting that nature's things make the laws to which they themselves are subject. But democratic rule is not to be understood here as leaving nature's things to their own devices. To the contrary, in Bacon's program, learning nature's laws fundamentally means disciplining nature by manipulative interrogational contrivances, contrivances that are designed to compel nature to confess its causal secrets. Such is the essence of experimentation. Bacon ultimately, in fact, proves to be no democrat at all. Having forced nature to yield the secrets of its inner workings, Bacon then envisions employing nature's laws in the complete subjugation of nature's things to human purposes. In Bacon's work people are meant to govern the realm of appearance and to control nature's things in order to suit their own needs.

Like Descartes, who followed him in time, Bacon turned his back on tradition and sought to break new ground for human understanding, and, most especially, for all knowledge claims concerning nature. Bacon expressed this aim metaphorically within the voyaging imagery of early-seventeenth-century exploration and discovery where he spoke his hope for the progress of science, as follows: "[I]t is fit that I publish and set forth those conjectures of mine which make hope in this matter reasonable, just as Columbus did, before that wonderful voyage across the Atlantic, when he gave reasons for his conviction that new lands and continents might be discovered besides those which were known before; which reasons, though rejected at first, were afterwards made good by experience, and were the causes and beginnings of great events."[33] This same imagery was employed in the frontispiece of Bacon's *Great Instauration*, which outlined his new science: there, the ship of learning passes between the Pillars of Hercules, heading outward from the known Mediterranean world through the Strait of Gibraltar, into the uncharted waters of the Atlantic.[34] Learning is bound for discovery of things not known. Indeed, learning itself will be constituted anew and become one with discovery.

Learning as discovery and the attendant imagery capture well Bacon's sense of launching a new science. And just as the point of a scientific discovery turns the mind to things outside of itself, so the compass Bacon will follow cannot be made out of the mind's own stuff in the manner of some Anglicized *cogito*. As Bacon views the matter by way of a pointed simile, the rational-deductive method to which Descartes later subscribed resembles the work of the spider that weaves a world out of its own substance.[35] Nature's things and what is worth discovering about them cannot, however, be drawn forth from logical-mathematical reasoning the way a web materializes out of a spider's interior. Nature in some way needs to be encountered on its own terms if we are to lay hold of what counts as real discovery, and the compass Bacon sometimes is said to have followed is that of inductive-empiricism, which contraposes his approach to Descartes'. Such a face-off makes for neat philosophical discussion and strikes the right contrasts between French and English intellectual tastes, rationalistic-deductive and empirical-inductive, respectively.[36] Yet treating Bacon as an empiricist risks obfuscating the very nature of his program of learning.

If we are Cartesian, we can immediately appreciate the force of the following: "[B]y far the greatest hindrance and aberration of the human understanding proceeds from the dullness, incompetence, and deceptions of the senses...."[37] But what sounds strikingly Cartesian is, in fact, Bacon's own assessment of our sensory apparatus. This judgment by no means makes Bacon Cartesian. As he envisions the acquisition of new knowledge, Bacon calls for the tools of induction. The material the stonemason Bacon uses to fashion a new foundation for knowledge nevertheless comes not from any empirical quarry. Whereas empirical thinking looks upon sense data as that-which-is-knowable-as-given, Bacon's thinking rests on the premise that "the testimony and information of the sense has reference always to man, not to the universe;

and it is a great error to assert that the sense is the measure of things."[38] The issue arising, then, concerns how Bacon avoids becoming a spider once he has renounced empiricism, and he confronts the issue directly as follows: "I said at the beginning and am ever urging, the human senses and understanding, weak as they are, are not to be deprived of their authority, but to be supplied with helps."[39]

"Helps" may sound odd to our ears, and its meaning tends in any case to be anemic because it lacks the clear and forceful thrust of what Bacon has in mind. By "helps" Bacon intends nothing less than experiments that, at their best, succeed in discovering nature's secrets. Unlike the *cogito*, the experiment stays in immediate touch with nature's things; experimentation entails "experimental observation," which is perhaps what leads many to call Bacon an empiricist. Speaking of experimental observation nevertheless misplaces emphasis by granting observation the status of noun while reducing experimentation to the grammatical function of a mere modifier.[40] As Bacon intends things, matters are the other way round: "[T]he sense by itself is a thing infirm and erring ... but all the truer kind of interpretation of nature is effected by instances and experiments fit and apposite; wherein the sense decides touching the experiment only, and the experiment touching the point in nature and the thing itself."[41] Or, as he reiterates this last point elsewhere, "I contrive that the office of the sense shall be only to judge of the experiment, and that the experiment itself shall judge of the thing."[42]

Bacon's point of interposing experimentation between our senses and nature's things would amount to nothing if the kind of experimentation he has in mind were to consist in simply monitoring nature, monitoring being a form of empiricism. Bacon does not discount experimental monitoring entirely or in every case, but he is primarily promoting a particular kind of experimentation that keeps in touch with nature's things in a profoundly singular regard. Bacon addresses this experimentation where he outlines his plan for the instauration of the sciences, based on what he calls "natural history":

> Of this reconstruction the foundation must be laid in natural history, and that of a new kind and gathered on a new principle.... For first, the object of the natural history which I propose is not so much to delight with variety of matter or to help with present use of experiments, as to give light to the discovery of causes and supply a suckling philosophy with its first food.... Next, with regard to the mass and composition of it: I mean it to be a history not only of nature free and at large (when she is left to her own course and does her work her own way) ... *but much more of nature under constraint and vexed* that is to say, *when by art and the hand of man she is forced out of her natural state, and squeezed and moulded*.... I do in fact (low and vulgar as men may think it) count more upon this part both for helps and safeguards than upon the other, seeing that *the nature of things betrays itself more readily under the vexations of art than in its natural freedom* [italics added].[43]

The experimentation Bacon here spotlights for modern science subjects nature to human artifice. What is *given* or presented experimentally to the senses and the faculty of understanding is not nature "left to her own course" where she can "do her work her own way." What is given, rather, is a nature "under constraint," "vexed," "squeezed," "moulded"—a nature mediated by experimentation itself. This kind of experimentation puts nature entirely at men's disposal: nature's things or their substitutes are to be manipulated in view of a design that intends to elicit from nature an answer to some question of knowledge. The particular question itself largely determines the manipulative design. In its design, the experiment typically combines conceptual, imaginative, and manual elements that pose a question nature might never have to answer if left to itself. What is interrogative then turns interrogational once the design is implemented as a set of procedural operations. Nature's things become subject to vexatious handling, or manhandling, that aims to wheedle or force some confession. Optimal success obtains when nature responds within the experiment in some revelatory way such that the secret sought after is discovered; which is to say, whatever we are looking for is experimentally disclosed to the experimental eye. When Bacon interposes experimentation between our senses and nature's things, he identifies the realm of appearance that modern science will aim to govern.

Appreciating how experimentation might constitute a governable realm of appearance preliminarily requires identifying the particular kind of secrets Bacon seeks to uncover. And he tells us in this regard that "my course and method, as I have often clearly stated and would wish to state again, is this ... from works and experiments to extract causes and axioms, and again from these causes and axioms new works and experiments, as a legitimate interpreter of nature."[44] Experiments are designed to extract causes and axioms: fixed, deterministic causes are the secrets of nature—the secrets of how nature works—that experimentation seeks to discover, and axioms are the causal laws at which repeated experiments and inductive reasoning ultimately arrive. The logical problem posed for induction by synthetic inference suffers obliteration of a practical sort when repeated experimental success satisfies pragmatic sensibility. From an experimental point of view, that is, certain knowledge of a particular natural law obtains when every properly conducted experimental test yields the same result: our knowledge works, and it works via experimental control! Knowing how causality works realizes certainty in the very act of mastering nature, just as Bacon tells us in the most famous of his aphorisms: "Human knowledge and human power meet in one; for where the cause is not known the effect cannot be produced. Nature to be commanded must be obeyed, and that which in contemplation is as the cause is in operation as the rule."[45] Sketching Bacon's point here in an ideal-typical experimental context will elucidate how such causal knowledge can and does govern the realm of appearance.[46]

Causal knowledge obtains where human ingenuity artificially meets with the cooperation of nature in an experimental context. At the outset, before we have such causal knowledge, we design an experiment to inquire into the causality of some phenomenon. Today we speak of such a specific question being posed to nature as a *hypothesis.* At one point Bacon tells us that experimentation cannot fail inasmuch as our question finds an answer whatever the experiment's outcome: the hypothesis is either right or wrong, and we have learned something in either case. So much may be true, of course, yet looking at the same matter from another angle, the experiment truly succeeds when its interrogational design operationally yields causal knowledge by way of nature's answering "yes" to the hypothesis. In this case we have discovered what causes the particular phenomenon. However, such discovery is provisional. To be sure of our discovery — to be certain — we need to corroborate the finding by repeated tests of the experiment. Assuming that repeated tests replicate the same result, we drop the provisional idea of knowledge being hypothetical and speak, instead, of possessing certain knowledge or of knowing the causal law. Any proper demonstration of our knowledge at this point directly reproduces what we already know; that is, the very replication of the experiment is itself a manipulative design by which we can certainly effect the appearance of the phenomenon. In the selfsame way that we are able to cause the phenomenon to appear, we command the realm of appearance. Mastery of nature arises foundationally at this experimental level of controlling causality and commanding appearance. Applied science or the utilization of knowledge for practical purposes principally rests and programmatically builds upon this "maker's knowledge."

Unlike Descartes' God, who realized Creation in the beginning with the Word, Bacon's God is a Maker. Bacon knew full well, of course, that in actual fact we are not God-the-Maker and that nature has something critical to say about what will or will not work. "For the chain of causes cannot by any force be loosed or broken, nor can nature be commanded except by being obeyed. And so those twin objects, human knowledge and human power, do really meet in one; and it is from ignorance of causes that operations fail."[47] Whereas Descartes' *cogito* fails to subordinate nature by rule of rational sovereignty alone, Bacon proposes to make nature follow her own rules under human command. Yet Bacon's knowledge and that of modern science do indeed constitute a "maker's knowledge," which Antonio Pérez-Ramos addresses in a manner germane to our purposes in his fine essay, "Bacon's forms and the maker's knowledge tradition." He writes, "Now there can be degrees of certainty in the way we proceed to describe the hidden workings of Nature. There can hardly be when we utilize a given description as a recipe for reproducing a phenomenon," and, as he later continues, "if a scientific statement can lead to the successful (re)production of the phenomenon it purports to describe ... then that very statement should be accepted as a full denizen into the realm of practically sanctioned knowledge."[48] Peter Dear makes the same point in his illuminating work *Revolutionizing the Sciences*, where he declares that, for

Bacon, "knowing *what* a thing is and knowing *how to produce* it are essentially identical."[49] In maker's knowledge, then, the experimental design becomes a recipe, a blueprint, for the manufacture of what is to appear, and nature appears upon instrumental command. We therefore speak of maker's knowledge here, as well as throughout this work as a whole, in express recognition of both meanings of *to make*: to produce and to compel.

Bacon distinguishes between two interests governing experimentation for different purposes. "Experiments of light" aim to discover nature's causal works in the interest of human understanding, and "experiments of fruit" aim to provide the mechanical arts with useful knowledge.[50] Of the two, Bacon holds the former to be the loftier. Yet we associate Bacon's name primarily with experiments of fruit, and for good reason. As he himself distinguishes his program from the old Aristotelian learning and medieval scholasticism, "the end which this science of mine proposes is the invention not of arguments but of arts; not of things in accordance with principles, but of principles themselves; not of probable reasons, but of designations and directions for works," and as he thrusts home the contrast, "[a]nd as the intention is different, so, accordingly, is the effect; the effect of the one being to overcome an opponent in argument, of the other to command nature in action."[51] The command Bacon has in mind finds full fruition in the mechanical arts, which Bacon showcases in fictional form in his *New Atlantis,* the fragment of a utopia published in 1627, a year after his death.

New Atlantis fictionally illuminates the marvelous material fruits of a science that has been socially organized to advance human benefit by way of applied knowledge. Experiments of light are important in this utopian world, but the enterprise and glory of science respect the provisioning of human need and social betterment. Social and political authority belong to the scientific elite, who are organized in the House of Solomon, which also is called the College of the Six Days' Works. Conjoining wisdom with the study of all Creation, the College legitimates its authority by delivering on its social purpose, which one of the House of Solomon's "fathers" articulates as follows: "The end of our foundation is the knowledge of causes, the secret motions of things; and the enlarging of the bounds of human empire to the effecting of all things possible."[52] We would be hard put to find any purer articulation of the project of mastery, which envisions nature being fully employed in the service of human need and desire under the rule of our scientific and technological dominion. Although but a literary fragment and scanting in its detail, *New Atlantis* presents an impressive array of the fruits of applied science, including a mechanical means of flight, underwater ships, scientifically improved sources of food, cures for disease and the scientific prolongation of human life, to name a few featured achievements. Bacon's utopian vision in this way showcases human domination of nature from the ground up, as it were. Nor is it mere coincidence that the first decidedly modern utopia envisions the project of mastering nature. Indeed, the project of mastery is what, in point of fact, defines this utopia's very modernity.

Bacon's prescience was altogether remarkable.[53] Just as manipulative-interrogational experimentation is the realm of appearance on which modern science chiefly came to rest its pragmatic claim to authority, so governance of that realm in lawfully causal knowledge largely legitimated the success of the scientific revolution. For all his prescience, however, Bacon's program for modern science proved just as deficient as Descartes,' though the other way round. Bacon failed to grasp the essential role mathematical objectification would play in marrying the experimental-instrumental domain of modern science to the theoretical-conceptual domain in which nature's laws might be understood and integrated. The theoretical-conceptual domain of modern science would, indeed, be mathematical through and through. Hence, whereas Descartes' ideas lacked hands for manipulating nature, Bacon's stood in need of mathematical vision. Each had an essential interest in commanding the realm of appearance and establishing certain knowledge, yet each one's program spotlighted but one substantial part of the new ground on which the scientific revolution established its success and erected a structure of knowledge. The successful marriage of the mathematical objectification of nature with manipulative experimentation in Isaac Newton's work fulfilled the early modern quest for certainty, and modern science was born neither Cartesian nor Baconian, but Newtonian. Moreover, in fulfilling that quest for certainty, Newtonian science meant the possession of truthful knowledge and the promise of dominion over this Earth for those who experienced enlightenment in the eighteenth century. It is to that experience that we now must turn.

2

Authorizing Human Dominion

The experimental and mathematical approaches to nature combining in the epistemological foundation of modern science were not legislated by Cartesian and Baconian fiat. Between the time of Copernicus and that of Newton those approaches found increasing favor in the cultural reorientation the West was experiencing, as well as in specific discoveries and achievements in astronomy and physics. Calling these historical developments between Copernicus and Newton the "scientific revolution" inevitably betrays some Whiggish sensibility. That is, from our perspective in this study, we read into the sixteenth and seventeenth centuries a progressive story — we plot and advance a meaning — that unfolds through the actions but behind the backs of historical actors who themselves possessed no such narrative understanding. The very terms *science* and *scientific*, for example, secrete a later historical meaning for historical figures who called themselves natural philosophers and understood their work to be in natural philosophy.[1] In exactly this same vein we speak ordinarily of World War I, yet no one fighting in that conflict had any such understanding. What we face here is a dilemma from which there is no escape, and science historian Steven Shapin admits as much even as he criticizes the Whiggish idea of there ever having been a "scientific revolution." He writes, "[T]here is inevitably something of 'us' in the stories we tell about the past. This is the historian's predicament, and it is foolish to think there is some method, however well intentioned, that can extricate us from this predicament."[2] Ongoing historical experience subjects earlier meaning to alteration necessarily, and it would be silly to deny hindsight its privilege when it comes to historical storytelling. Without getting into historiographical issues lying far outside the scope of this work, then, in speaking of there having been a scientific revolution we simply hold with Hegel that any real understanding of experience is like the owl of Minerva, which "spreads its wings only with the falling of the dusk."[3]

As Minerva's owl would have it, Copernicus and Newton are typically the bookend figures in the various historical treatments of the "scientific revolution."[4] In the beginning, as it were, Copernicus offered a model of planetary motion that expressly contradicted the long-accepted Ptolemaic astronomy and directly subverted the Christian-Aristotelian cosmology explaining it. And, in the beginning, Copernicus' heliostatic model was accepted for employment as merely a mathematical calculating device having no claim to truth. By the time of Newton's work 150 years later, however, matters had changed fundamentally. In the end Newton's work established the physical reality of the heliostatic picture by way of a scientific explanation having purely mathematical form and experimental support, and such a mathematical reality was proclaimed without concern for the authority of institutionalized religion. Indeed, Newton himself quickly was apotheosized, and within a generation of his death in 1727 his work achieved biblical status throughout the Occidental world. And what scripture that work proved to be! Truth on Earth also meant the possibility of human dominion.

This chapter initially will look very briefly at select aspects of the scientific revolution from which Newton's work sprang. From that point we then will explore how Newtonian science took hold of European understanding, at least the understanding of the cultured, educated elites. We will argue that the very way the first Newtonians disseminated the new natural philosophy meant both certain truth and the promise of mastery to those bearing witness to the knowledge claims being made. Bearing witness meant, more particularly, observing humankind's mastery of nature's workings in the practical demonstrations of knowledge being carried out both in experimental exercises and in machine employments by those spreading the new understanding. And to those first generations of Europeans living within spoken memory of the cultural anarchy that had beset an earlier time, such witnessing most fundamentally meant secular revelation. This new natural philosophy, Newtonian science, seemed to deliver humankind from the benighted nature of all understanding existing theretofore! It was an absolutely new beginning — the genesis of real understanding that brought with it real dominion over the fish and fowl of the Earth while throwing light upon all Creation. Or so people believed without doubt.

The eighteenth-century Enlightenment, known also as the "Age of Reason," arose with this epiphanic sensibility. The dramatic impact of truth revealed nevertheless receded experientially as the century ran its course, and the truthfulness of Newton's explanation of how reality works quite soon simply was taken for granted. On a practical plane the proof of such truth plainly appeared in the mastery of nature exercised experimentally whenever and wherever Newtonian science advanced knowledge and discovery. And scientific progress at once underwrote the larger idea of progress which came to possess the eighteenth-century imagination: social and material progress surely would spring from the apparent fact that knowledge and power truly do meet in one.

For educated Westerners, the force of this Baconian aphorism now had behind it the unquestionable authority of modern science.

Looking into the promise of human dominion in this chapter will take us further than this cultural interpretation, however, as we will elucidate further the nature of the master's knowledge in direct light of Newtonian science. The mastery we saw in maker's knowledge in the preceding chapter will reappear here, but now command of the realm of appearance will be one with the pragmatic authority of Newtonian science. And in that authority, in the nature of the causal determinism of its understanding, we will learn something very important about the timelessness of the master's knowledge.

Glimpsing the Scientific Revolution

Integral to the gestation of modern science during the century and a half separating Copernicus and Newton were the growing emphasis on mechanical crafts and experimentation on the one hand, and the changing appreciation of what mathematics might mean and achieve on the other. The increasing emphasis given to mechanics, technology, and experimentation, or to instrumental knowledge in general, manifested a significant cultural reorientation that had been underway for some time.[5] Classical culture and medieval culture alike had privileged the *vita contemplativa* (contemplative life) over the *vita activa* (active life) and in so doing depreciated along with worldly work any instrumental modes of knowledge.[6] Bacon's call for a new program of knowledge to replace Aristotle's sought to reverse this invidious distinction. His call for works in the place of words and experimental knowledge in the place of rhetorical logic programmatically focused an already widening interest in instrumental understanding that sprang from the developing navigational needs of voyaging, the growing attention to mining and metallurgy, and from various alchemical traditions such as that promulgated by the influential Paracelsus in the sixteenth century.[7]

Along this same line, practical instruments proliferated in the seventeenth century, powerfully extending the reach of our senses in "passive instruments" like the microscope and telescope and broadly expanding our experimental operations with "active instruments" like the air pump and the condensing engine.[8] Most of these influences and inventions, though not all, made their way outside of university curricula, many in non-academic institutions featuring experimental investigations of nature. The most notable of these institutions were the Academy of Lynxes in Tuscany, which included Galileo, the Royal Academy of Sciences in Paris, and the Royal Society of London, which expressly identified its mission with Francis Bacon's vision and which included Isaac Newton among its many seventeenth-century luminaries.[9] But if instrumental knowledge was beginning to claim its due outside of universities, universities themselves did play a critical role in the changing employment of mathematical reason, which was undergoing some basic revaluation at this very same historical juncture.[10]

Medieval university learning had consisted mainly of scholasticism, which applied reason to divine revelation, as one scholar has put it.[11] Mathematics and geometry possessed second-class citizenship in the scholastic curriculum. Allowed no consort with cosmology, Ptolemy's geometry served merely to picture God's Creation and to calculate celestial occurrences like eclipses. Ptolemy's astronomy certainly enabled people to predict those occurrences, but ancient Babylonians also had used mathematics to make similar predictions with a very different cosmology in mind. Medieval thinking looked upon mathematics as a mere tool; it fit a calculative function and nothing more. Not everyone in early modern time concurred, however, and by the mid- to late seventeenth century new thinking was well in place. Out of the Renaissance had come an appreciation of ancient Greek thinking that would help provide new employment for mathematics.[12] Plato, Pythagoras, and Archimedes were being heard, and what gained an ear, of course, is the idea that mathematics might have something to do with how nature is put together.

These Greek influences proved seminal in preparing late medieval and early modern thought for the shift to a mathematical understanding of nature, and in the work of Kepler and Galileo these influences were especially significant. In the larger cultural context, however, the shift showed a remarkably Cartesian character. Speaking of the seventeenth century most generally, Richard Westfall puts the matter this way: "Only a few followed the full rigor of the Cartesian metaphysic, but virtually every scientist of importance in the second half of the century accepted as beyond question the dualism of body and spirit. The physical nature of modern science had been born."[13] University curricula in this regard were particularly important in objectifying nature via Descartes' physics. Descartes' physics proved faulty in a measure not unrelated to his failure to grasp the essential significance of experimentation, but as John Gascoigne points out, "By the late seventeenth century ... universities throughout Europe were actively disseminating Cartesian natural philosophy and, in the process, were helping to introduce the educated classes to the major conclusions of the mechanical philosophy."[14] In general, then, as the spreading appreciation of instrumental knowledge worked its way mainly along extracurricular paths, the mathematical treatment of nature worked its way largely through university channels featuring Descartes' work.

Within this general historical framework individual achievements were most significant. Newton declared that if he had seen farther than others, it was because he stood on the shoulders of giants. He may have had in mind classical titans such as Aristotle rather than more proximal figures like Descartes, but along with Descartes were other early modern giants whose work blazed the mathematical and experimental trails successfully converging in Newton's work. Johannes Kepler drew upon the vast welter of Tycho Brahe's astronomical data in fashioning the three laws of orbital motion that replaced Copernicus' circular motion with ellipses, eliminating thereby the visual clutter of the Copernican model with mathematically precise visual simplicity.[15] For new understanding to emerge, however, astronomy needed physics, and explanation of the Copernican-Keplerian universe found much of its material here on Earth in the study of local motion. That domain of study was Galileo Galilei's. Although Galileo is known primarily for his telescopic discoveries and his legendary run-in with the declining authority of the Roman church, he was much more essentially the founder of modern physics. His studies of motion established in temporal terms the measure of a material body's freefall, and in establishing the magnitude of velocity as a function of time, Galileo opened nature to the investigation of force. Once force came into the picture, astronomy moved beyond pure geometry, making heavenly motion subject to physical dynamics. Quite simply, Galileo's mechanical wedding of mathematical reasoning and experimentation enabled him, at least in principle, to bring the heavens within his experimental reach.[16]

Unable to handle planets or any other celestial body, humans nevertheless can handle physical things directly in the investigation of local motion. Balls can be lifted and dropped, or rolled along incline planes, or put into trajectory by cannon. So, too, the study of ballistics can plot the motion of things subject to the concurrent forces of inertia and gravity, capturing thereby the reality of planetary motion here on Earth. And just as terrestrial things might be seen as proxies for celestial bodies, so too the study of bare bodies moving in space and time bore with it the universality of nature's laws.[17] But Galileo's thinking on this score came up a bit short at one crucial conceptual point. Turning his back on the Aristotelian understanding privileging rest as the physical state towards which all motion aims, Galileo idealized the conditions of motion and conceived of a body moving interminably without external resistance. He thus advanced thinking about dynamics towards the law of inertia, putting himself seemingly in place to discover in the parabola, which describes the motions of planets and cannonballs operating under the same combined forces, the isomorphism unifying heaven and Earth under the same physical law.[18] But Galileo kept full faith with Copernicus' commitment to the circle, and wrongly conceptualized inertia as entailing circular motion.[19] As Dudley Shapere puts the point, "In contrast to the extension of terrestrial physics to the astronomical realm Newton was to accomplish, Galileo ... made terrestrial physics a branch of astronomy, the study of circular motion."[20]

Galileo's mathematical-experimental physics just missed claiming modern science in his name, but his work nevertheless delivered the study of dynamics to the doorstep of modern science. And with that work in place, the company of early modern giants stood ready to support Newton on their shoulders. Rupert Hall makes these basic connections as follows:

> By distinguishing the real from the apparent motion of falling bodies Galileo removed a principal objection against the Copernican system; by discerning the beautiful simplicity of elliptical orbits behind the apparent complexity of the planetary revolutions Kepler increased the likelihood of its truth. Newton's achievement was founded on theirs, for by perfecting the Galilean mechanics he found the basis for a final distinction between reality and appearance that placed Kepler's astronomy beyond doubt. The two great strands of seventeenth-century physical science united in Newton's *Principia* (1687) through the fertile and fitting combination of mathematics with the mechanical philosophy that yielded the laws of gravity, when Kepler's planetary laws were shown to be but special cases of the laws of motion foreshadowed by Galileo.[21]

And so we come to Isaac Newton, whose work assumed truly biblical status not only for those educated elites in the West who received it, but for generations to follow both in the West and beyond.

Experiencing Secular Revelation

The irony of Newton's achievement is that what would be called his scientific work constituted for him but a small part of a far more profound metaphysical quest for religious understanding. All kinds of esoterica hold hands with science in Newton's wider work in religion and alchemy, and John Maynard Keynes spoke insightfully when he called Newton the last of the great magicians.[22] Yet Lord Keynes' judgment most likely surprised more than a few of his early-twentieth-century readers because the entire modern age had effectively suppressed what did not conform to the Newton who belonged to science. Only Newton's scientific achievement had counted, and what did not seem to suit that achievement simply did not matter. While we recognize the distortion such partiality entails for a biographical understanding of Newton or an appreciation of the corpus of his work, it is nevertheless precisely that achievement and that achievement alone that interests us here.

Speaking of Newton's achievement in the singular intends no slight to his paternity rights to countless accomplishments in science and mathematics. We mean only to spotlight in very large terms why for nearly two centuries Newtonian laws governed the understanding of nature either in point of fact, as in physics and astronomy, or as a model for most other fields of inquiry. Giants there may have been, yet it was the fate of Newton's genius to unify the heavens and Earth in lexical magnitudes and laws of motion in what he variously called "experimental philosophy" and "rational mechanics." Experimental philosophy and rational mechanics meant one and the same thing, though it turned out that Newton featured the different adjectives separately in his two major works. In the magisterial *Principia* Newton unfolded a mathematical treatise emphasizing the *rational* in rational mechanics, and in his exemplary *Opticks* he presented a treatise consisting of applications of *exper-*

imental philosophy. The two works together constitute the pillars of Newton's scientific work, and while the *Opticks* became the experimental guide in many emergent branches of eighteenth-century science, it was the terribly recondite *Philosophiae Naturalis Principia Mathematics* (1686), to give the *Principia* its full title, that spelled out the causal laws governing the motions of bodies in space and time.

In the very first of the *Principia*'s "Rules of Reasoning in Philosophy" Newton declares, "We are to admit no more causes of natural things than such as are both true and sufficient to explain their appearances."[23] The mechanical causality underlying appearances reduces to forceful contact between material bodies: not the forceful contact pictured in the medieval impetus theory, which we might imagine in Sisyphus' constant application of his shoulder to the boulder, but the forceful contact we find later pictured in the classical billiard ball analogy that captures material contact and impressed force in dynamical fashion. Newtonian physics is an impact physics, and while Newton himself was no atomist, his impact physics invited the atomic picture of physical reality that Newtonians came to see lying behind all appearances. From the very outset Newton's work promised to grasp nature's reality wherever one might want to explore nature's things. Roger Cotes directly stated this promise in his "Preface" to the second edition of the *Principia*, which Newton had asked him to write: "Fair and equal judges will therefore give sentence in favor of this most excellent method of philosophy, which is founded on experiments and observations.... The gates are now set open, and by the passage he [Newton] has revealed we may freely enter into the knowledge of the hidden secrets and wonders of natural things."[24]

Cotes' exultation sounds essentially Baconian, and Newtonian science certainly would discover and entail the recipes of maker's knowledge. But Newton worked differently than Bacon had. Newton's experimentation entailed a mathematical grasp of nature's causal mechanisms. Theory and experimentation interpenetrated. Newton puts the matter directly in a private letter where he wrote:

> I told you that the theory which I propounded was evinced to me, not by inferring *'tis thus because not otherwise*, that is, not by deducing it only from a confutation or contrary suppositions, but by deriving it from experiments concluding positively and directly. The way therefore to examine it is by considering whether the experiments which I propound do prove those parts of the theory to which they are applied, or by prosecuting other experiments which the theory may suggest for its examination.[25]

In Newton's work, the recipes of maker's knowledge implied natural laws just as natural laws implied recipes. But more immediately, within the historical context of the scientific revolution, Newton's astrophysics successfully united heaven and Earth in the mechanical laws governing all motion when he thrust his law of universal gravitation into that particular breach between reality and appearance opened so dramatically by Copernicus. His law of gravity did not close the breach: it simply demonstrated the causality behind the appearance of celestial motion and thereby corrected the misimpression of our senses, which had misled humankind since time immemorial. But what an achievement that was, for with it, Isaac Newton gave people scientific reason to believe in the mechanical laws explaining the Copernican-Keplerian universe.

Newton's work was not recognized all at once by everyone everywhere, but it did win out among competing belief systems before too long. Some people, of course, never accepted Newton's work, and those who truly counted among the naysayers usually had their own versions of a mechanical reality, either Cartesian or other. There did, moreover, exist amidst several lacunae in Newton's work one glaringly ironic issue of major magnitude: Newton offered no mechanical explanation of gravity itself. Gravity was the causal reality explaining planetary motion, the Earth's included, yet concerning the apparently empty space through which gravity operated, Newton seemed to allow for no mechanism or dynamical cause.[26] Newton's lawful universe was anything but complete or fully consistent on its own premises. Our present interest lies, however, not with any deficiency in Newton's understanding, which we will entertain in the next chapter. The sole interest here is in Newton's achievement, which in its cultural impact and practical success blunted the point of any and all critical objections to his rational mechanics. That cultural impact amounted to a very great deal indeed, and it found quite clear and precise articulation almost immediately in the telling words of Newton's contemporary, the poet Alexander Pope:

> Nature and Nature's Laws lay hid in Night:
> GOD said *Let Newton be*! and all was Light.[27]

Possessing a poetic feel for the pulse of his time, Pope seems to have understood the magnitude of Newton's achievement in the movement of Western historical experience between early modern time and the modern age. Certain knowledge indeed! Alexander Pope was experiencing secular revelation!

Nietzsche tells us in the parable of the madman that human deeds are like the sound of thunder and the light of stars; they need time for people to hear of them and to see them for what they are. So it was with Newton's work, which needed the time of one to two generations to take complete hold of the Western world. England embraced its own quite quickly, but Newton had to make his way on the Continent, which took longer in some places than in others. By the mid-eighteenth century Newtonian science had in any event become virtually synonymous with understanding itself: directly, as regards physical nature, and indirectly, as a model for most other fields of study. Secular revelation generally meant that without scripture or miracle, human beings had established truthful understanding—certain knowledge—of things by employing their hands experimentally and using their reason mathematically.

The experimental manipulation of things and the objectification of nature were happily married.

The actual dissemination of Newtonian thought did not proceed either uniformly or in unalloyed fashion, and in many cases what came to be called "Newtonian" was neither pure nor accurate.[28] Various French rationalists received Newton as a pure rationalist while many of the practical-minded Dutch tended to see Newton simply as an experimenter. Such culturally different intellectual propensities also suggest local distilleries making different blends. In France, for example, it was hardly extraordinary to see Newtonian and Cartesian physics combined. Given such blending, anything at all mechanistic might loosely be called Newtonian, giving one historian of science, Thomas Hankins, reason to argue that the term "Newtonian" really constitutes nothing specifically meaningful apart from an eighteenth-century ideology promulgated by the philosophers and widely accepted in popular belief. Hankins dismisses "Newtonian" as useless to an historian of science, and though he may be right in particular instances, his judgment surely should not be categorical.[29] In and of itself the very existence of such an ideology is telling of what Newton achieved since it demonstrates the reach of secular revelation in Newton's name.

Newton's England lost no time in acclaiming their own as the immortal whose work had unlocked truth and utility for all humankind. Given the Baconian bent of English sensibility and Newton's affiliation with the Royal Society, his success took root in home soil immediately. At the very outset of the eighteenth century Newtonians John Keill and Francis Hauksbee publicly employed experiments to prove Newton's mathematical demonstrations, and with the appointment of Jean Desaguliers as Curator of Experiments for the Royal Society in 1714, this public approach became programmatic with the further addition of machine demonstrations.[30] In England and the Low Countries alike, Desaguliers offered numerous series of public lectures directly promoting the marriage of the new natural philosophy to the mechanical arts. "Desaguliers became the most famous of literally hundreds of eighteenth-century Newtonian practitioners who applied the mechanical legacy of the *Principia* to a wide range of practical problems, from the draining of mines and swamps to the building of canals and electrical experimentation."[31] This approach also found favor in school curricula in Cambridge and, perhaps more influentially, in academies governed by non-Anglican Dissenters. As Dobbs and Jacob tell us of the impact these schools had in England by the end of the century, "[by then] any self-respecting business family with industrial interests opted first to send their adolescent boys to such an academy, or possibly to Edinburgh University, where medical and scientific education was among the best to be found in Europe."[32] The alliance of science with business and industry was in its mere infancy in England in the second half of the eighteenth century, but it is most noteworthy that the industrial revolution began just then and there.

England would not become the school of Europe, however, at least not in the same respect that Pericles had once called Athens the school of all Hellas. In the case of the new natural philosophy, "it was the conversion of key Dutch universities—the most popular institutions of higher learning in Europe—that made Newtonian prosyletizing efforts a success."[33] Beginning in Leiden University with Burchand de Volder's conversion from Cartesian to Newtonian understanding, a succession of powerful and extremely influential Dutch lecturers proclaimed the truth of Newtonian science, experimentally demonstrated its utility, and promoted its application to different fields of inquiry. Hermann Boerhaave attained widespread fame across two or three generations of students from all over Europe, first and foremost as a lecturer in medicine but then later as a professor of botany and chemistry as well. Yet medicine remained his niche, and, according to historian Peter Gay, "Boerhaave taught medical Newtonianism.... His textbooks, promptly and widely translated, were models of Newtonian reasoning; they acquired almost scriptural authority and reached those who could not come to hear him."[34] Alongside Boerhaave, Willem Jacob van 'sGravesande lectured in mathematics and astronomy but promoted Newton's experimental physics as well in his public demonstrations and publications; 'sGravesande's texts sometimes turned the mathematical treatment of problems in physics into tracts of civil engineering, and he himself occasionally served as an engineering consultant on public projects in the Netherlands.[35] This same approach was taken by Petrus van Musschenbroek, a student of 'sGravsande's, who became an author of leading textbooks on rational mechanics and experimental philosophy. And another Musschenbroek, Petrus' brother Jan, became widely known for fashioning precision instruments that served scientific experimenters throughout Europe, his brother and 'sGravesande included. Those instruments needed ideas to find full employment, of course, and just as students from far and wide might carry Newtonianism away from lecture rooms and public demonstrations held in the Netherlands, so translations of the Dutch textbooks into various European languages helped to spread Newtonian ideas all over the Continent. Not in France, however, as France proved to be a very special case.

Nowhere was Cartesian thought more deeply entrenched than in France, for obvious reasons. The rising tide of Newtonian thinking during the opening decades of the eighteenth century was hardly likely to change that, and some French Cartesians actually looked upon Newtonian natural philosophy as a "national malady" of the English.[36] But even France ultimately yielded, though it took until mid-century. Intriguing in the French case is that the success Newton's work came to enjoy there did not strictly follow the paths cut by the professional mathematicians, experimenters, mechanics or natural philosophers in general, as was the case almost everywhere else. In France Newtonian science enjoyed the support of a literary figure who had converted to Newtonianism and whose wide fame stood Newton's work on a stage where it could be seen and discussed in glowing public light. This convert was Voltaire, who, like many who convert to something, embraced his subject with missionary zeal. Voltaire's own conversion took a little time, beginning in 1728–29 with his visit to England, where he fell in love with a political order and cultural world far less restrictive than that in France. Having been deeply impressed by Newtonian science during that visit, Voltaire later announced his defection from all things French in his *Philosophical Letters on the English* (1733). There he gave considerable praise to Bacon, Locke, and Newton, and in so doing he offered the French their first real taste of things Newtonian. In that work Voltaire spoke directly of Descartes and Newton together, at one point declaring, "I indeed believe that very few will presume to compare his [Descartes'] philosophy in any respect with that of Sir Isaac Newton. The former is an essay, the latter a masterpiece."[37] Voltaire's *Letters* culminated but the first phase of his affair with Newton, for having consulted with 'sGravesande and Pierre de Maupertuis, France's leading Newtonian, Voltaire then undertook serious study of the Englishman's work with the aim of making Newton's thought publicly accessible. What followed was Voltaire's *Elements of the Philosophy of Newton* (1738), which, despite its mathematics, reached lay audiences as well as professional ones, and which, even in the eyes of contemporaries, helped to eventuate a Newtonian "revolution" in France.[38]

Voltaire's work joined with that of legions of professionals in spreading Newtonianism, and through such efforts, the scientific revolution actually became a cultural revolution in the way Europeans were being educated and coming to see things in an altogether new light. Dobbs and Jacob write:

> By the middle of the eighteenth century, the extent and range of European domination enjoyed by Newtonian science, both practical and conceptual, had become wider and more sophisticated than even Newton — master experimenter as well as conceptualizer of space and time —could ever have imagined.... By the middle of the eighteenth century European children of some affluence were being taught to think mechanically while their parents, first men and then gradually women, were attending lectures in Newtonianism.[39]

This process of diffusion brought with it Newton's apotheosis, which is hardly surprising. Voltaire himself called Newton the greatest human being who had ever lived. Having witnessed Newton's funeral without seeing any resurrection, Voltaire was assured of Newton's mortality; yet as Mordecai Feingold writes, "The religious metaphors deployed by Voltaire are seminal to his conception of Newtonian science and to his awareness of his evangelical mission. Newtonianism became something of a secular religion for Voltaire...."[40] Voltaire's own commitment to spreading the gospel certainly was in keeping with the larger cultural faith to which it was contributing. Voltaire's was the century of the philosophes, and where Peter Gay speaks generally of that "Age of Reason," he tells us directly, "In the deification of Newton, the Enlightenment of the philosophes and the age of Enlightenment were at one."[41]

Such deification was not the work of natural philosophers, instructional experimenters and Voltaire alone. Even while Newton still lived poets were immortalizing him in all the major European languages; and throughout the eighteenth century and across the Continent, biographers and painters alike helped craft his secular divinity.[42] Behind it all lay the striking force of Newton's achievement so precisely captured in Pope's couplet: throughout the ages Nature and its Laws had lain hid in Night, but once God gave Newton to humankind, all was Light. The religious feel of secular *revelation* so apparent here underlies the deification of Newton, and there is no question that the religiosity of the early eighteenth century lives meaningfully within God's letting Newton be and, accordingly, Newton's bringing the Light. This religiosity also must have played some role in the faith people initially had in modern science. But the revelation was secular, and we need to be crystal clear about the idea of secular revelation so as not to have our meaning connected to, much less confused with, that of Carl Becker, whose *Heavenly City of the Eighteenth-Century Philosophers* (1932) fathered a line of interpretations of the Enlightenment to which the present interpretation does not subscribe. To address Becker's understanding in a nutshell, we need only to substitute the deification of reason for the apotheosis of Newton.

Becker sees the Enlightenment's faith in reason cast in a medieval religious form, such that the progress envisioned in Bacon's *New Atlantis* constituted heavenly salvation, and faith in scientific reason differed in no essential way from faith in scriptural truth. The new priesthood were the eighteenth-century philosophes—those literary and philosophical men of letters like Voltaire, Diderot, Hume, Montequieu, et al., who typically contraposed science to organized religion and promoted progress in the faithful name of reason and science. Becker puts the matter this way:

> [I]f we examine the foundations of their faith, we find that at every turn the *Philosophes* betray their debt to medieval thought without being aware of it.... They renounced the authority of the church and Bible, but exhibited a naïve faith in the authority of nature and reason.... They denied that miracles ever happened, but believed in the perfectibility of the human race.... In spite of their rationalism ... there is more Christian philosophy in the writings of the *Philosophes* than has yet been dreamt of in our histories.... [Indeed], the *Philosophes* demolished the Heavenly City of St. Augustine only to rebuild it with more up-to-date materials.[43]

In the same way secular *revelation* bears witness to a religious sensibility in the early eighteenth century, Becker is right to speak of a secular faith. And inasmuch as many philosophes kept that faith with three pictures in their studies—one of Newton, Bacon, and Locke each, so we might speak of a trinitarian secular sensibility, which is most ironic given Newton's own Arian heresy.[44] But where Becker and his followers go wrong, or at least where we depart from any such company, is with the functional equivalence they assume between reason and religion, science and scripture, or philosophes and priests.

Becker's interpretation is reductionistic to the point where no really meaningful difference obtains between a medieval monk's faithful belief in scripture and Voltaire's faithful belief in Newtonian science.[45] The authority of scripture in no small part rests upon witness to miracles, but those writing scripture were themselves secondary sources quite removed from the venues where such unnatural events were alleged to have occurred. Moreover, miracles were, and are, especially hard to accept millennia later in an age that began with reasonable cause to distrust appearances in the very first place. David Hume's critique of miracles remains a signature piece of the Enlightenment's perspective on belief in such events.[46] Newtonian science, on the other hand, rested its own authority on what men could do and could see right in front of each other, giving common witness or testimony to events that were anything but unnatural. In the place of miracles, scientific proof entails not only mathematical demonstration but, much more convincingly, the demonstrations of experimental knowledge. Scientific "witnessing" in such contexts begins with predicting what will occur and then causing what has been predicted to occur in fact. Witnessing of this sort is a far cry indeed from whatever religious thaumaturgy may involve.

Grand corroboration of Newtonian science came in the mid-eighteenth century with dramatic proofs of three predictions following directly from Newton's work: measurements taken during Maupertuis' polar expedition confirmed the oblong shape of Earth; mathematical work on the three-body problem favorably resolved the difficulty of measuring lunar motion; and Halley's comet returned exactly when it was supposed to, so to speak.[47] Such testimony, which entailed nothing of a miraculous character, was of the stunning sort that makes headlines. Yet it was the more prosaic evidence of nature's obedience to the Newtonian laws, laws that were of nature's own making, of course, that carried the day among the students and publics witnessing the experimental and machine demonstrations carried on by men whose work we have discussed. By mid-century Hauksbee, Desaguliers, and 'sGravesande, along with many others, had already spread the word through countless proofs that nature operated according to causal laws that Newtonian science could explain and control in its instrumental knowledge. The faith of those bearing witness was as real and as reasonable as those demonstrations of scientific knowledge were themselves convincing. Indeed, while their sense of revelation may have expressed the sensibility of a religious era fast disappearing, those experiencing *secular* revelation were at the same time witnessing scientists' command of nature in the exercise of maker's knowledge.

Celebrating Mastery in Progress

A great many people of the eighteenth-century Enlightenment fast became "oh so certain" of their reason, of human capabilities in general, and of the certain prospect, if not the inevitability, of continuing human progress. Far removed they surely were from the quest for certainty dominating the early part of the preceding century. They were soon to become "oh so certain" of their own abundant self-sufficiency that, in point of fact, at the table where such secular celebration carried on, God's very place became insecure. For men like Kepler and Galileo, the Creator Himself had appeared in the harmonic geometry of motion and in the mathematical character of nature's form and action. And for Descartes, the founder of modern rationalism, God had truly been his very first certainty, which led Descartes later to make pilgrimage to Italy to give thanks to God for having allowed him to discover the *cogito* in himself. Giants cutting the trail to modern science, those men were, nevertheless, very much of the seventeenth century, as was Newton, whose intensely personal Christianity claimed far more of his time than did any of his work in rational mechanics. But as Dobbs and Jacob speak to this last point, "faith was not ultimately the use to which Newton's system was put. In a sense the world had passed Newton by in his own lifetime, and [his] concerns with religion and with finding a unified system that encompassed both nature and divine principles ... were of much less concern in the eighteenth century."[48] Thinking changed in the eighteenth century.

Mainstream religious sensibility among educated circles in the eighteenth century followed the channel cut by mechanistic thinking and critical reason. Early in the century this sensibility primarily found articulation in "deism," which actually had begun in England in the seventeenth century and which rapidly spread throughout the culture of the philosophes.[49] Deists typically rejected organized Christianity and revealed religion. Not for them the myriad sects, cannibalistic ritual, Biblical mythology, or belief in miracles. In the face of dogma and occasional fanaticism deists campaigned for religious toleration and justice, and behind their mechanical universe they saw the hand of a master craftsman, a divine maker, who had practiced rational mechanics in His six days' work. Such a God lost His immediacy and became remote, which made smooth if not logical the transition from deism to the atheism so evident in the later eighteenth century. After secular revelation had captured European belief in the first half of the eighteenth century, materialism became increasingly pronounced, and with it grew the reach of atheism. The haughty could then become quite brash indeed, as was evident in Laplace's reply to Napoleon. Having just published a full explication of the Newtonian universe, which his own mathematical studies had helped to perfect as a mechanical system, Laplace had to face a critical query Napoleon himself put to him after he had read Laplace's work. In what was otherwise a favorable response, the political demigod registered dismay at the fact that nowhere in his scientific account of the universe had Laplace once mentioned God: to which the author famously replied, "I have no need of that hypothesis."

The certainty manifest in Laplace's reply was brash in many ways. He was, after all, addressing Napoleon Bonaparte. But the problem God Himself had with finding some place in the celebration of secular revelation really had less to do with any assertive atheism, though there was plenty of that, than with a growing religious indifference sprung from the secular enlightenment of the time. Historian Peter Gay puts the matter this way: "In the first half of the century, the leading philosophes had been deists and had used the vocabulary of natural law; in the second half, the leaders were atheists and used the vocabulary of utility."[50] Turning attention away from things beyond this world, modern science and its instrumental promise increasingly focused people's eyes on the material world and what could be done with it.

What had for so long been God's Creation now became for a good many people simply a material world of things. Turning to utility, men and women preoccupied themselves with their immediate needs and desires relative to the perishable things of this world. It was not God who came to mind in these concerns, but Francis Bacon. Bacon's maker's knowledge was widely evident in the machine demonstrations and experimental proofs men like Desaguliers and 'sGravesande had performed in the promulgation of Newtonianism, and as modern science took increasing hold of people's consciousness, thinking turned more and more to applied science, which was well known to be Bacon's bailiwick. In lauding Bacon along with Locke and Newton, the eighteenth century paid homage to "the first great statesman of science," to employ Loren Eiseley's phrase,[51] and saw in the utopian possibilities Bacon had envisioned, matters falling within the promising reach of their practical knowledge. The expanding success of experimental philosophy in widening spheres of application pragmatically endorsed the prospect of human dominion over nature's things, and in such endorsement secular revelation and mastery of nature—truth and utility—came to one and the same thing. Recalling Bacon's aphorism, human knowledge and human power truly were meeting in one. They were meeting in the idea of progress, in fact, which initially had referred to expanding knowledge, but which came to be associated largely with material utility as the eighteenth century unfolded. In short, material progress really is what the practical thrust of scientific knowledge pointed towards in the popular imagination of the modern age.

Devotees of science and progress met for faithful assembly not in churches, but in salons, lodges, coffee houses, and at public lectures and demonstrations where talk of the good and the true turned not to scripture but to utility. J. B. Bury's classic judgment that "Physical science was important only in so far as it could help social science and administer to the needs of man"[52] fails utterly to appreciate in secular revelation the very spirit and self-consciousness of the eighteenth-century Enlightenment. It fails also to realize that scientists would continue seeking truth as distinct from utility. Yet Bury does speak knowingly of the popular interest in science. Interest in the social utility of knowledge was evident everywhere, most especially in the widely read and historically renowned *Encyclopedia*, which is the defining document of the age. This work would become the model for all encyclopedias to follow, as was intended by Denis Diderot, whose guiding hand and early editorial association with Jean le Rond d'Alembert steered the twenty-year project through financial and political storms, which twice included governmental suppression of its publication.[53] Consisting of twenty-eight volumes, seventeen of text and eleven of plates, publication ran from 1751 to 1772 and drew from contributors of all sorts throughout Europe. Anything knowable "from A to Z" might find its way into print, but the project's spirit and direction expressly followed the utilitarian compass provided by Francis Bacon.

Writing the entry "encyclopedia" for the *Encyclopedia*, Diderot articulated the utility of the work itself as a foundation for the collection and development of knowledge:

> In truth, the aim of an encyclopedia is to collect all the knowledge scattered over the face of the earth, to present its general outlines and structure to the men with whom we live, and to transmit this to those who will come after us, so that the work of the past centuries may be useful to the following centuries, that our children by becoming more educated, may at the same time become more virtuous and happier, and that we may not die without having deserved well of the human race....[54]

Interesting here is the way Diderot intended the *Encyclopedia* itself to help realize progress, and to this day, of course, the supplements that are so much a necessary part of the use of encyclopedias bear witness in the short term to this idea of the continuing advancement of learning.[55] In Hankins' words, the Enlightenment's "*Encyclopedie* was to be a new beginning, a foundation of knowledge, whose improvement by succeeding generations would ensure human progress...."[56] And it was Bacon whom Diderot and d'Alembert had in mind in their editorial enterprise. Planting their tree of knowledge in Baconian soil, they followed Bacon's tripartite scheme for organizing knowledge according to the faculties of memory, reason, and imagination. Then, in his "Introduction" to the first volume of the work, d'Alembert proudly placed Bacon at the head of those "illustrious personages," including Descartes and Newton, whose work had prepared the light of reason which then illuminated the world.[57] Diderot's very choice of how to educate himself for lead editorship of the project clearly foreshadowed the Baconian bent of the entire enterprise, for he visited the workshops of craftsmen in and around Paris, interviewing the craftsmen themselves and taking notes on the ways of their maker's knowledge.[58]

The *Encyclopedia* applauded the eighteenth century as the age of philosophy, but the philosophy it honored had nothing to do with the *vita contemplativa.* To the very contrary, nothing prejudicial towards the mechanical arts was permissible, as d'Alembert made quite clear in a statement reflecting as much the temper of the age as the intellectual propensity of the work he was helping to edit. He wrote,

> [T]he advantage which the liberal arts [intellectual arts] have over the mechanical, because the former demands hard, intellectual work and requires difficulty to excel, is sufficiently compensated by the far greater usefulness the latter arts for the most part provide for us. It is their very utility which forced these arts to be reduced to merely mechanical operations, so that a greater number of men could practice them. But society, in justly respecting the great geniuses which have enlightened it, need not on that account vilify the hands of those who serve it. The discovery of the compass is no less advantageous to the human race than the explanation of the properties of the compass needle is to physics. Finally, considering in itself the distinction we are discussing, how many of the so-called scholars are there for whom science is, in reality, only a mechanical art?[59]

This multi-volume work spanning twenty years would not so much proselytize in the name of the new rational sensibility as display that sensibility in material developments and knowledge claims of greater or lesser practical substance. Quite simply, the *Encyclopedia* exhibited progress, and the exhibition attained its most impressive heights in the new material achievements and mechanical processes that were showcased "from A to Z" in the eleven volumes of plates. As Franklin Baumer tells us, along with instructional words there came hundreds of pictures "depicting the techniques of the trades and crafts of contemporary Europe — a sort of review of technology — on the eve of the Industrial Revolution. Following Bacon here, Diderot envisaged a society in which the mechanical arts, allied with science, would give power over nature for man's benefit."[60] The *Encyclopedia* was the key cultural work of the eighteenth-century Enlightenment because, in substantiating the material foundation of progress and encouraging critical reason in all spheres of human endeavor, it brought to public attention the instrumental intersection of scientific knowledge, technological control, and social benefit. That intersection defines the project of mastery.

The Enlightenment feasted there in that intersection where the project of mastery was both present, as on report, and promissory, as in the making. For some feasting in the material celebration, progress seemed most promising when reason extended beyond nature's things and brought society or particular social institutions within its reach. Rational criticism of the forms and practices of existing institutions flowed daily in the conversations, letters, and essays of the philosophes, who typically intended to spark reform in one way or another. In some cases this intention to influence or launch rational socio-political reform assumed large proportions in major social studies texts attaining wide readership and gaining significant recognition at the time. The greatest of these were Baron de Montesquieu's *Spirit of the Laws* (1748) and Adam Smith's *Wealth of Nations* (1776), both of which transformed the historical tissue of their material into social science.[61] Montesquieu's work, which became instantly paradigmatic in its "scientific approach," founded political sociology, and Smith's book established the science of economics before that science became "dismal." Each work aimed to disclose lawful mechanisms underlying and explaining social phenomena, and, respecting those lawful mechanisms, each work intended enlightened reform to follow. Smith tendentiously wrote to persuade governments to abandon mercantile policies in view of the material progress he saw issuing forth inevitably from the rational causality of the market mechanism, which, left to itself, operated according to the laws of competition. Of course, Montesquieu's formulation of the separation of powers famously provided the "Founding Fathers" of the United States Constitution a political mechanism to safeguard the people and their republic against tyrannical power. As Ernst Cassirer more generally informs us about Montesquieu, "Montesquieu is a man of his time, a genuine thinker of the Enlightenment, in that he expects from the advancement of knowledge a new moral order and new orientation of the political and social history of man.... From a knowledge of the general principles and moving forces of history he looks for the possibility of their effective control in the future."[62]

Social science emerged with the hope that men might rationally control social development, and in the eighteenth century this hope led some thinkers to envision progress in the extreme. Joseph Priestley saw the matter this way:

> *[K]nowledge*, as Lord Bacon observes, being *power*, the human powers will, in fact, be enlarged; nature, including both its materials, and its laws, will be more at our command; men will make their situation in this world abundantly more easy and comfortable; they will probably prolong their existence in it, and will grow daily more happy, each in himself, and more able (and, I believe, more disposed) to communicate happiness to others. Thus, whatever was the beginning of this world, the end will be glorious and paradisaical, beyond what our imagination can now conceive.[63]

Priestley here predicates the open-ended possibilities of progress on the basis of science and its advancement of material civilization, and progress becomes boundless beyond our mastery of nature. The Marquis de Condorcet expresses this same utopian sensibility, basing his vision upon mathematics and physical science but arguing much further "that no bounds have been fixed to the improvement of the human faculties; that the perfectibility of man is absolutely indefinite; [and] that the progress of this perfectibility ... has no other limit than the duration of the globe upon which nature placed us."[64] Extreme beliefs like these once led the historian Crane Brinton to call progress and perfectibility the "corollary doctrines" coordinating eighteenth-century beliefs.[65]

Thinking like Condorcet's stretched the meaning of human progress quite far beyond the purview of the natural philosophy giving rise to it, and Brinton's judgment may itself be somewhat extreme. Belief in human perfectibility would hardly have obtained anything near a majority vote among the philosophes. That said, the French Revolution (1789–1794), which ironically took Condorcet's life, nevertheless did stretch the meaning of progress and reason beyond all sensible limits. Among other things, the French Revolution was a political affair of men and women who addressed their entire world as an object of rational design and maker's knowledge. Using America's constitution-making as a springboard for ambitions far more profound, the French Revolution culminated the furthest extension of the project of mastery outside the natural world — Reason disclosing the Truth of how society is to be properly made over; Reason treating all institutions as artifacts that, along with time and space, might be altogether redesigned and controlled anew.[66] But at this point we must draw back and recall the idea of mastery from the verge of historical narratives pointing too far away from the scientific revolution and its impact on the West.

The presumption underlying all such extensions of reason and progress was the belief that human beings either do possess or could possess knowledge that would enable them to remake and master the world. The kind of knowledge lying at the heart of this belief in itself had nothing to do with any revolutionary actor's rhetoric or any scholarly treatise in social science. Maker's knowledge and its great promise pertained, rather, to the relationship between men and things culminating in Newtonian science, where it really did appear that knowledge and power meet in our understanding of nature. Having already discussed in the preceding chapter how causality might govern the realm of appearance in maker's knowledge, let us now reconsider such knowledge in the context of Newtonian science. And we shall begin by looking at the nature of the authority modern science brought with it.

The Master's Knowledge

A historian readily can find connections between Renaissance sensibility and Enlightenment thinking, but those eras manifested very different approaches to the Western heritage. Renaissance thinkers had revered the classics, finding among the ancients models of excellence in all domains of human endeavor. Eighteenth-century intellectuals tended, on the other hand, to depreciate and in some cases directly disparage classical thought. Such critical judgment extended beyond matters of natural philosophy, though there it certainly was unqualified. Enlightenment thought directly inherited the position taken by the "moderns" in the so-called quarrel between the ancients and the moderns arising at the end of the seventeenth century. In that quarrel the moderns had inverted the relationship of authority between past and present theretofore characteristic of Western thought and upheld most vehemently during the Renaissance. Whereas for the so-called "ancients" in the quarrel the Greeks and Romans were superior elders whose authority rested upon their unsurpassable excellence in experiential wisdom and cultural accomplishments, for both the moderns and the philosophes who followed, the ancient Greeks and Romans were in fact children whose ignorance and shortcomings sprang from their lack of experience and their less sophisticated sensibilities. Immanuel Kant, arguably the eighteenth century's greatest thinker, in this vein gave no quarter whatsoever to the past when articulating the motto of the Enlightenment. Wanting to realize the self-determination of reason in freedom from external authority — that of the past as well as all other heteronomy — Kant answers the question "What is Enlightenment?" with the following declaration: "Enlightenment is man's release from his self-imposed tutelage. Tutelage is man's inability to make use of his understanding without direction from another.... *Sapere Aude*! 'Have the courage to use your own reason!' — that is the motto of the enlightenment."[67]

The eighteenth century did not abandon the past altogether, of course. Gibbon wrote his monumental work on Rome; Hume turned from philosophy to the *History of England*, in six volumes, no less; and classical poets and writers still possessed avid audiences. Yet history, though it may not have suffered total eclipse, nonetheless did seem to many simply to lack meaningful credentials *qua* history. The idea that history possesses meaning on its own terms held diminishing credibility, and when the past did find meaning, its significance often lay in service to the future. This subordination of history was implicit in Diderot's *Encyclopedia*. Those philosophes who founded social science worked in a similar fashion, for however real their historical understanding may have been, they placed their historical material in the service of "science" as they sought to discover the laws behind social action and institutional forms; laws that might, in turn, serve social reform in the interest of enlightened progress.

This depreciation of the past climaxed in the radical politics of the French Revolution. The French Revolution rocked the entire Western world, giving birth to modern socialism and nationalism, as well as defining the political lexicon of our time. Kant's motto became identified with bloody nihilism in the minds of men like Edmund Burke, who founded European conservatism in response to the French revolutionaries' extirpation of the *ancien regime*. Attacking the "Republic of Reason" and its thoroughgoing assault on the past, Burke argued against the belief that traditions and institutions are geometrical entities subject to contractual reconstruction by rational design. He argued that society "is indeed a contract ... a partnership not only between those who are living, but between those who are living, those who are dead, and those who are yet to be born...."[68] Such a contract had been found mortal, however, a full century or more before the Revolution. Indeed, the abandonment of the past — or, more precisely, the abandonment of the past-as-tradition — occurred with the rise of modern science, the new natural philosophy.[69]

Developing the new natural philosophy had meant the wholesale rejection of the philosophical-theological traditions underlying the Christian-Aristotelian cosmology. And *new* this incipient modern science surely was, as so many early modern works themselves titularly attest: Bacon's *New Organon* and *New Atlantis*, Galileo's *Discourses and Demonstrations Concerning Two*

New Sciences, Kepler's *New Astronomy*, and the list goes on.[70] We earlier saw in a different context how establishing new foundations for knowledge had entailed jettisoning the past, leaving the old texts behind and breaking new and different ground. Considered here in the present context, that foundational separation of modern science from the authority of the past succeeded only because, as things turned out, the new natural philosophy was able to purchase authority entirely on its own terms without inheriting, or needing to inherit, any legacy of the past. Modern science gained this success and established its own authority throughout the Western world by the middle of the eighteenth century, and in fulfilling the early modern quest for certainty, science constituted itself as a new, modern "tradition."

Calling the authority of science a new tradition in the first instance means understanding that Newtonian science filled the cultural void created by the breakdown of scriptural tradition. Science now authorized knowledge of nature, or of God's Creation. Calling the authority of science a new "tradition" also means understanding that modern science constitutes a new kind of cultural tradition, one that in fact turns the idea of tradition completely ironic, if not essentially meaningless. Prior to the new scientific understanding, all cultural authority going back to ancient Roman time bound itself to traditions that were themselves institutionally sedimented in time. Authority abided in the privileges accorded long-standing beliefs and long-established cultural practices. Authoritative traditions, therefore, were those possessing cultural roots and historical duration, and authority itself was quite literally *time-tested*. Modern scientific authority, on the other hand, at its core had little or nothing to do with temporal matters. Indeed, the very success of Newton's experimental philosophy shattered the previous interrelationship of time, tradition, and authority.

The success of Newtonian science fundamentally lay in the combination of pragmatic testing, where maker's knowledge can operate experimentally and in machine demonstrations, and causal understanding, where the theoretical formulation of laws obtains mathematically. Men like Desaguliers and 'sGravesande relied on no authority external to the practical efficacy of the Newtonian laws. As Dobbs and Jacob speak of the *Principia*, "In addition to its grand vision of the universe, it provided a rational mechanics for the operations of machines on earth. The artisans ... had never before had definite physical laws to enable them to calculate and predict mechanical forces and rationalize their control."[71] Now they did! Mastery of nature in this respect provided modern science its own pragmatic measure of success. The fact that Newtonian science *worked* placed the authority of science beyond all the claims of the past, beyond any test of time. That authority simply relied on the timelessness of nature's laws, which scientific knowledge could compel to operate instrumentally.

There is another aspect of the timelessness of the scientific master's knowledge, in fact a crucial aspect of such knowledge, that we need to see apart from the larger cultural landscape we have been addressing thus far. In speaking of timelessness in this respect of maker's knowledge, we intend to understand how the master's knowledge actually abandons time in the exercise of dominion. Our discussion of maker's knowledge in the first chapter will help us here, with an added bonus: having Newtonian science in hand, we are no longer without a scientific understanding of nature, as we were in discussing Bacon. Whereas Bacon had had only a general methodological compass to guide him beyond the Pillars of Hercules, Newtonian scientists had the *Principia* in hand as they set out to explore nature with a lawful understanding of nature's ways. And at just this juncture Morris Kline offers us a wonderfully large window on the material at hand:

> Over and above the identification of the Earth and the heavenly bodies, Galileo's and Newton's work established the existence of universal, mathematical laws. These laws described both the behavior of a speck of dust and the most distant star. No corner of the universe was outside their range.... The work of Copernicus, Kepler, Galileo, and Newton made possible the realization of many dreams. There was the dream and hope of ancient and medieval astrologers of anticipating nature's ways. There was also the plan that Bacon and Descartes had advanced of mastering nature for the improvement of human welfare. Man progressed toward both goals, the scientific and the technological. The universal laws certainly made possible the prediction of the phenomena they comprised. And mastery is but a step away from prediction, for knowing the unfailing course of nature makes possible the employment of nature in engineering devices.[72]

Kline's historical appreciation is illuminating, but it is his analytical understanding of mastery as but a step away from prediction that hits the present mark dead center.

Possessing scientific knowledge meant knowing what would occur without question; it meant certainty. So much was true because Newtonian science realized knowledge upon causal laws that were predictive and certain. The earlier "dream of anticipating the ways of nature" therefore found Newtonian substance in strict causal determinism. Laplace, providing more than one quotable quote, articulated the classical statement of this strict, causally predictive determinism with the same bold confidence he had shown Napoleon. Imagine, Laplace urges, a superhuman Newtonian intellect:

> An intellect which at a given instant knew all the forces acting in nature and the position of all things of which the world consists—supposing said intellect were vast enough to subject those data to analysis—would embrace in the same formula the motions of the greatest bodies in the universe and those of the slightest atoms; nothing would be uncertain for it, and the future, like the past, would be present to its eyes.[73]

This universe possesses a mechanical perfection that Newton himself had not seen, but this universe operates strictly in accordance with the Newtonian laws and is entirely predictable on that basis.[74]

Following the conventional shorthand classically identifying mechanical causality, $f = ma$, where the quantity of force is equal to the product of mass times acceleration, what the Laplacean seer needs to know are the positions and momenta of all masses in the universe at a given instant. What this picture would amount to is a colossal description of state. A description of state is like a snapshot photo of a physical system, which in the Laplacean case pictures the motions of all bodies in their spatial and temporal magnitudes at the present moment in time. Laplace's interest, of course, is not to dream Faust's dream of knowing all things, but to identify the nature of the predictive certainty of Newtonian knowledge. Inasmuch as all things are present to the eyes of the Laplacean seer, his vision really has less in common with Faust than with the Augustinian God, who sees all that has ever been and all that ever will be inasmuch as He exists outside of time. And this resemblance is telling indeed, not because either a real or an ideal Newtonian scientist could live outside of time, but because the nature of strict causal determinism privileges the Laplacean seer's knowledge in a manner transcending time. In such causal knowledge time has no meaning: neither the past nor the future has any relevance whatsoever. This point wants clear precision so as to foreclose any possible misunderstanding, for the knowledge we are considering here is critical to our understanding in general. The knowledge we are addressing here belongs to the master.

The atemporal character of the master's causal knowledge in no sense denies the fact that nature's causality is physically embedded in time: duration abides in the sequential reality of cause and effect, where effect follows cause in time. Prediction itself observes such temporal reality in its very nature: prediction pre-dicts; it speaks before the actual event, which will occur later. So too, of course, flesh and blood scientists live in the everyday world where space and time coordinate our every experience of things, and scientists' continual striving after new discoveries and scientific progress surely witnesses the significance of the future. But none of these realities is implicated in the understanding at hand here, where time plays no role in the causal knowledge under discussion. Recalling Laplace's point, for such an all-knowing Newtonian seer "nothing would be uncertain ... and the future, like the past, would be present to its eyes." In the strict mathematical determinism of Newtonian laws, causal knowledge collapses time within a description of state. Time disappears in the snapshot description of a physical state as if somehow an entire film might be captured and seen within a single frame.

Some years ago Milič Čapek approached this same point from a purely philosophical angle. Time suffers a logical flattening within the deterministic framework of the knowledge we are considering. There is within that framework no difference between our moving from the present to the future, which is cause to effect, and our moving from the present to the past, which is effect to cause. Our knowledge going in one direction is logically symmetrical with knowledge going in the other direction, leading Čapek to observe, "[A]ccording to classical mechanics the difference is comparable to the conventional difference between a plus and a minus sign in analytical geometry" and then to ask, altogether rhetorically, "But if the asymmetry of time is abolished, is not time itself eliminated?"[75] This very same elimination of time is real but suppressed in the position taken by so many philosophers of science who speak of the "logical symmetry between explanation and prediction." According to this idea, which neatly accords with the hypothetico-deductive model of science, if we can explain a phenomenon, moving from effect to cause, then we are equally able to predict it, moving from cause to effect.[76] In a later chapter we will return to this point of the presumed symmetry between explanation and prediction in scientific knowledge, but here we need to consider how such timelessness "works" in experimentation.

This timelessness works experimentally in the same way any demonstration of our scientific knowledge of a strictly causal law differs from the pursuit of knowledge via some causal hypothesis. We discussed this difference in the preceding chapter in connection with Bacon's program of mastery, and here we need simply to reiterate that difference in terms of time. The provisional — or uncertain — status of an hypothesis requires the experimenter testing the hypothesis to remain on top of the experiment, being closely watchful and following up the experiment fully in order to determine whether or not the hypothesized cause does actually produce the predicted effect. The hypothesis may prove to be wrong, in which case the experiment will not produce what was predicted. Experimentation therefore requires learning what happens in fact, and respecting whatever may or may not happen, time is absolutely telling in the operation of such experimentation. Time is not telling, however, when an hypothesis has gained the status of knowledge through repeated tests that all replicate, or reproduce, the predicted effect. In this case, any experimental demonstration of certain knowledge turns the experimenter into a kind of practicing Laplacean seer who knows in and from the contrivance of the given, which corresponds to a description of state, exactly what will follow. While the laboratory can be as tricky and messy as the world outside of the laboratory, as any science teacher will tell you, and ordinarily requires continuing vigilance on the experimenter's part, in the ideal typical experiment under discussion here, the key is for the experimenter simply to "get it right" and then immediately get out of nature's way. Unlike the human slave who needs watching to insure that the work order be carried out, nature has no choice but to follow its own rules, which in this case are made to work by experimental command. The experimenter's knowledge is not time-dependent. The experimenter's knowledge is independent of time precisely because the experiment itself is a contrivance manipulatively designed to produce the predicted effect through instrumental command of the causal law being demonstrated. Maker's knowledge produces as it compels, and certainty means being independent of time.

Control over the realm of appearance is one with knowledge of the causal law governing what will occur experimentally, and applying such knowledge in the larger spheres of social utility takes the short step from (scientific) prediction to (technological) mastery of which Morris Kline spoke. Be it a matter of basic experimental research or of applied science, the shorthand form for such commanding knowledge finds classical expression in the formula "*if this is given, then that will occur.*" This simple "if ... then" formula captures the Laplacean seer's perspective: it captures his certain knowledge of strictly deterministic causality as well as the timelessness of his commanding knowledge. In short, the "if ... then" formula captures the essence of maker's knowledge; it belongs to the master.

The timelessness of the master's knowledge only superficially resembles the timelessness of the truths belonging to the *vita contemplativa.* Athens and Jerusalem handed down eternal verities that largely survive today either as matters of personal faith, religious or philosophical, or as cultural artifacts of abandoned convictions. Scientific laws of the kind we have been discussing rest their timelessness on firmer foundations. Newtonian laws themselves are timeless in their instrumental efficacy exactly because they are of nature's making. Human dominion needs consist in following nature's recipes as Bacon well knew, and with the rise of Newtonian science the promise of such dominion became one with progress, which modern science appeared absolutely to guarantee on its own authority. The full thrust of the project of mastery did not arrive, however, until the nineteenth century when the industrial revolution appeared to confirm in ubiquitous material fact the cultural belief that we moderns are master makers of our world. The industrial revolution materialized the project of mastery triumphantly for everyone to see the world over, and in so doing it deepened the conviction that modern science authorizes the project of mastery. This conviction is the legacy of Newtonian science; it is the cultural legacy of the story of modern science that we have told thus far. Yet not all is said and done in the story of Newtonian science or its cultural legacy. Turning to the nineteenth and twentieth centuries, we need to see both within and alongside the industrial revolution how Newtonian science would reach farther and farther into nature, far beyond the realm of material things that Newton and the early modern natural philosophers had managed to control. But tracking these matters and seeing where they lead takes our story beyond the present chapter. Having done with the rise of modern science and the origin of our belief that science authorizes human dominion, the owl of Minerva remains in flight.

3

Ironies of Scientific Progress

Nineteenth-century Europe was a time of revolution and bourgeois ascendancy. Many of those making revolution sought to remake society, pure and simple, while many of the bourgeoisie made profit in their daily business of making commodities. It was a maker's world, and out of that maker's world came a world made over. Stretching from the time of Napoleon into *La Belle Epoque*, the nineteenth century went from cart transport, kerosene lanterns, and horseback travel, to vast railway networks, electrified urbanscapes, and airplanes.[1] Made over indeed! What the eighteenth century had anticipated, the nineteenth century realized in and through the industrial revolution. The industrial revolution materialized wholly, though without plan, the progress promised, foreseen, and begun piecemeal during the preceding century. Throughout the Enlightenment science and technology had consorted with one another quite frequently, and in the case of James Watt, sometimes with significant offspring. But the wholesale and systematic application of knowledge to the production of things was conceivable only in some instrumental union of the two, which took place in their nineteenth-century industrial marriage and which has continued in various alliances ever since.

Nineteenth-century science did not lose its identity in that industrial marriage. Yet in serving the industrial revolution and benefiting from it, science did begin to develop the very characteristics marking it thereafter and defining it today. Science increasingly became specialized, professionalized, and organized around laboratory research requiring substantial funding. Nor did physical science, or, more particularly, physics, stand in place. Physicists pursued all sorts of phenomena under the rule of rational mechanics, and as the century wore on, scientists studying light, electricity, and magnetism found their paths converging both experimentally and theoretically in ways unexpected. Along those same pathways the investigation of nature became increasingly instrumentalized, not least because the very phenomena being investigated — electricity and magnetism — could be "handled" only by way of tools and apparatus. These phenomena could not be picked up by human hands, much less studied in depth, without artifice intervening at every level of experimental inquiry. Science and technology thus found deep bonding in the experimental realm as well as in industry. Following this development of physics will take us into the first decades of the twentieth century, where, in those deeper recesses of nature, Newtonian science proves inadequate in its understanding of physical reality, and then, in a truly revolutionary development, nature first announces its independence from experimental command. The legacy of Newtonian science would live on in the continuing belief that science authorizes our dominion over nature, but from the middle third of the twentieth century forward, such belief would in actual fact no longer be authorized by modern science itself.

This chapter initially will highlight key relationships between science and technology in the industrialization of Europe and then sketch the development of physics, first on the electromagnetic frontier of Newtonian science, and then in the subatomic realm. Following the development of physical science during the nineteenth and early twentieth centuries to the work of Einstein and, more profoundly, to the revolution wrought by quantum physics, this chapter will spell out a dual irony in the meaning of scientific progress. With respect to the historical development of physics, this irony consists in the fact that the very advance of scientific knowledge generally would prove fatal to the claims to truth and mastery marking the rise of modern science and establishing its pedigree. This irony of scientific progress at the same time reproduces itself culturally in a second irony, as it were. For at the very time when modern science was beginning to lose its commanding grip on nature's causality, the industrialization of society appeared everywhere to materialize in fact the progress-through-mastery visualized by Bacon in the seventeenth century and promised by science in the eighteenth.

The Early Alliance of Science and Technology

On May Day in 1851 Queen Victoria of England and Prince Albert, her husband, together attended an event that made the day the *greatest* in English history, or so the young queen later would write her uncle, King Leopold of Belgium.[2] That event was the opening of the Great Exhibition of London at the Crystal Palace, a building consisting of steel and some 300,000 panes of glass of uniform size and, at one-third of a mile in length, enclosing more space than did most royal palaces of the time. During the six months of the Exhibition more than six million visitors passed through the doors of the Crystal Palace, marveling at the wonders of British industry and technology while bearing witness more generally to the burgeoning achievements of an increasingly material civilization. Historian Simon Schama observes that at that historical moment, which is when the word "Victorian" became part of the English language, the meaning of the word *industry* did a "semantic somersault," leaving behind its premodern denotation of labor expended to assume its modern denotation of labor saved.[3] The industrial cast of the myriad inventions, gadgets, and materials unmistakably showcased the way of the world and the shape of things to come: this was the industrial revolution trumpeting its success in a triumphant spectacle. The event might be British and commercial to a large extent, but similar spectacles would follow outside of Britain. Moreover, beyond nation and commerce the meaning of things would be clear to everyone in all such instances: the industrial revolution materialized human progress.[4] It materialized the very same human progress that had been first glimpsed by Francis Bacon, later authoritatively grounded in Newtonian science, and then promisingly displayed in the *Encyclopedia.*

The industrial revolution remade the West during the nineteenth and twentieth centuries, and the Crystal Palace is perhaps but its earliest emblematic moment, the moment when all eyes were on the eventful appearance of that "alliance between science and technology" that Hans Jonas says marks the industrial revolution's harnessing of our knowledge of nature to the art of invention in the very way anticipated by Leonardo and Bacon.[5] Instrumental knowledge found celebration in the Crystal Palace, just as the Crystal Palace made industrial progress the cynosure of Britain's claim to leadership among the nations of the West. Critics of instrumental values and material progress were on the scene as well, but their criticisms mainly targeted the loss of craft sensibility and the reduction of aesthetics to functional design in the products of industrial manufacture.[6] On the political front, the European Left provided numerous perspectives generally attacking the interests of capitalism lying behind the products, processes, and progress of industrialization. Yet even amidst the strongest criticism of capitalism so pronounced on the Left, recognition of capitalism's advancing instrumental knowledge and material progress sounded loudly and clearly:

> [Capitalism's] rule of scarcely one hundred years, has created more massive and more colossal productive forces than have all preceding generations together. Subjection of Nature's forces to man, machinery, application of chemistry to industry and agriculture, steam-navigation, railways, electric telegraphs, clearing of whole continents for cultivation, canalization of rivers, whole populations conjured out of the ground—what earlier century had even a presentiment that such productive forces slumbered in the lap of social labour?[7]

Speaking here of the bourgeoisie and their continuing conquest of nature, Karl Marx penned this praise in what is perhaps the finest encomium capitalism ever received in fewer than one hundred words. Capitalism's greatest critic looked through the relations of production, which he saw institutionalizing class exploitation, to the mastery of nature he saw necessary for transcending the conditions of economic scarcity and eliminating the grounds of human inequality. For Marx, capitalism's forces of production entailed the application of science to the processes of manufacture and to what today we would call projects in research and development. Those modern forces of production forged that "alliance" between our knowledge of nature and the art of invention Jonas identifies less with capitalism than with the industrial revolution in general. But however one conceptualizes that alliance institutionally, what remains certain about the intersection of science and technology in the nineteenth century is the materialization of the *vita activa*'s victory over the *vita contemplativa*—a victory Marx himself rhetorically sought to conjure in his "Eleventh Thesis on Feuerbach" where he famously declared, "The philosophers have only *interpreted* the world, in various ways; the point, however, is to *change* it."[8] Or, as Peter Dear concludes his fine work on European knowledge and its ambitions in early modern time, looking ahead from the scientific revolution to the modern world that would follow:

> European learned culture, in regard at least to an understanding of the natural world, had undergone a shift from a stress on the *vita contemplativa* ... to a stress on the *vita activa*.... "Knowing *how*" was now starting to become as important as "knowing *why*." In the course of time, those two things would become ever more similar, as Europe learned more about the world in order to command it. The modern world is much like the world envisaged by Francis Bacon.[9]

The resemblance highlighted here refers not to any imagined contemporary social organization of science bearing the likeness of the House of Solomon, much less to any fantasized utopian character of the world we inhabit. The Baconian allusion regards, rather, the increasing programmatic connection of science to practical purposes in the nineteenth and twentieth centuries and, with this, the application of maker's knowledge to the overall shape and character of our world. Not until the nineteenth century did maker's knowledge truly break free from the traditional cultural norms that long had retarded the uses to which practical knowledge could be put. Values binding manufacture to craftsmanship and the fear of the human capacity for evil survived well into the eighteenth century, serving still as barriers to the unleashing of instrumental potential already evident in the scientific knowledge of the time.[10] Mary Shelley's *Frankenstein* (1818) articulates such fear from the Romantic perspective, which generally saw much to fear in the face of scientific enlightenment and advancing industrialization. Yet in the first half of the nineteenth century such fear already was in clear retreat. Indeed, the very figure of Faust tells the tale. Born of medieval Christian culture in the mold of the sorcerer's apprentice, Faust makes his pact with the devil and brings destruction upon himself in the late sixteenth-century work of Christopher Marlowe. Two centuries later, however, the Faust reported to have been in G.E. Lessing's missing manuscript gains salvation by following Kant's enlightenment motto.[11] Cultural norms had changed, certainly by the end of the eighteenth century: unlike the transcendent values restricting technological activities under the reign of the *vita contemplativa*, the governing values of the *vita activa* stressed instrumental norms respecting means-ends relations, pragmatic standards, material progress, and the like.

The complete ascendancy of the *vita activa* in the nineteenth century's alliance between science and technology had been clearly augured in the preceding century or two. The technological application of modern scientific knowledge had been apparent at the outset in the work of men like Desagulier and 'sGravesande, who promoted the cause of Newtonian science in the service of various practical needs in mining and land use. Science found part-time employment in the spread of secular revelation, and such piecemeal and random applications of scientific knowledge or principles of scientific investigation increased throughout the eighteenth-century Enlightenment. Diderot and his team chronicled such intersections of science and technology, and the doctrine of progress accordingly elevated the old notion of *homo faber*— man the maker — to the rank of a master, whose scientific capabilities promised some meeting of Faust and Prometheus in modern time.

The previous chapter traced out this cultural impact of science and showed how during the eighteenth century the thrust of science took on an increasingly technological cast in the popular understanding of the age. Nineteenth-century developments decidedly reinforced such understanding as the quest for certainty and secular revelation receded even further from memory and truth-on-earth found itself increasingly confined to matters of utility. Yet technology and science are not altogether the same since the former defines itself strictly in the application of useful knowledge for practical purposes whereas science distinguishes itself at least partly in pure science, or, more directly, in the investigation of nature undertaken in the pursuit of understanding. The industrial revolution might blur this distinction to some extent, and in our own time the distinction comes even closer to obliteration, but the idea of a nineteenth-century alliance between science and technology assumes the distinction just drawn, and, indeed, industrialization progressed by way of ongoing fertile exchange between advances in the investigation of nature and continuing technological invention and commercial innovation. This idea of interbreeding captures that century's progress much better than does any conflation of science and technology.

Some fields of science actually created new industries out of wholly new theoretical insights and experimental successes, insights and successes born solely from the desire to understand natural phenomena. In such cases, rather than knowledge following upon some research interest in technological advance and/or industrial production, scientific knowledge itself gave birth to a range of technologies and industrial products unimaginable up to that point. This impact of science nowhere is more apparent than in the fields of electricity and chemistry, two fields which caused the operational relationship between science and technology to turn a new corner.[12] An old saw has it that necessity is the mother of invention. This notion is not altogether misleading, but it is another matter whether or not those inventions themselves led to their own diffusion in technological innovation. Innovation had historically almost always been a question of some felt need, and necessity had truly been the mother of diffusion and innovation, if not of invention. This situation changed dramatically in the nineteenth century when the newly emergent scientific fields of electromagnetism and chemistry made available instrumental knowledge that in and of itself created large and unprecedented possibilities of invention and practical application. As Jonas tells us, these scientific fields of investigation "all by themselves *originated* their own, novel arts of large-scale utilization. The respective technologies, springing as it were from Jupiter's head ... were thus the first *wholly science-generated* (and henceforward, science-guided) technologies in the history of mankind."[13]

Electricity is not anything someone can hold or directly handle. In fact, until modern scientific investigation discovered electromagnetism and brought electricity within instrumental reach, humans had little or no evidence of such phenomena. Lightning quite clearly had always been observable during fleeting instances, and people were long aware of frictional electricity without having any clear notion of what it was. What we now identify as electricity, however, meant nothing whatsoever to people before science entered upon the scene. In Jonas' words, "electricity is an abstract entity, disembodied, immaterial, unseen; and to all practical intents, viz., as a manipulable force, it is entirely an artificial creation of man."[14] What this fact means, of course, is that all electrical industries rest upon a phenomenon that scientific knowledge first had to produce out of itself, so to speak. Friction historically had been the only way humans could produce electricity, but such static electricity, known as far back as the ancient Greeks, had little practical value.[15] Not until the beginning of the nineteenth century was electricity meaningfully made available in the production of the battery, the important point here being that the battery itself was initially constructed as a means of producing electricity. In other words, scientific knowledge had to embody itself in a particular technology in order to bring into being, in order to *make*, a phenomenon that humans had not been able to appropriate directly from nature. In the science of electromagnetism, then, our knowledge of nature and the art of invention combined immediately in the very instrumentalization of our ability to investigate nature. And upon that foundation alone there followed a dazzling array of technological wizardries that would change the character of Western societies. Lewis Mumford writes:

> At this juncture, for the first time, scientific experiments in physics ... resulted almost within a single generation in inventions that derived almost nothing from an earlier technology: the electric telegraph, the dynamo, the electric motor; and within two generations came the electric lamp, the telephone, the wireless telegraph, and the X-ray. All of these inventions were not merely impracticable, but technically inconceivable, until pure scientific research had made them live possibilities.[16]

The same scientific elevation of human artifice was displayed somewhat differently but no less significantly in the development of chemistry, which founded a host of booming nineteenth-century industries upon new grounds of scientific knowledge and instrumental capability. These grounds were chemical synthetics that in and of themselves made possible modern pharmaceuticals and modern dye manufacture, among numerous other industries. Modern synthetics lay in a long line of artificial products, indeed in a long and illustrious history of manufacture embracing the ancient invention of bronze as well as the creation of steel. But unlike all such alloys of earlier human artifice, modern synthetics did not merely combine materials already provided by nature. In the case of chemical synthetics, scientific knowledge penetrated the constitution of objects by way of molecular manipulation, creating things lying beyond what nature itself imagined within the material of its own creations. Modern chemistry in unprecedented fashion thus enabled human beings to change the very substance of nature.[17] Such knowledge, such *making*, had little to do with the tradition of synthetic manufacture, or even with Victor Frankenstein's creation, which simply recombined materials already given. The synthetic capability we are addressing here points, rather, toward the recombinant DNA of our own time.

The science of chemistry also fathered institutional development that served as a model for much of the scientific research to be carried out in the industrial revolution. Chemistry pioneered the research laboratory that soon came to dominate the institutional nexus of science and technology, both in the university and in all sorts of commercial enterprises. As a venue for bringing together scientific discovery and technological application, the research laboratory became the locus of "the invention of the method of invention," which is the way Alfred North Whitehead once characterized the greatest invention of the nineteenth century.[18] Taking the lead on the educational front of applied science was the German system of higher education, which rapidly emerged during the first half of the nineteenth century as the leading force in the study of chemistry.[19] German universities and technological schools early on set the pattern that other countries emulated as the century unfolded, the most celebrated examples of such copying probably being the Cavendish Laboratory established at Cambridge University in England and Thomas Edison's industrial laboratory established at Menlo Park in New Jersey. For some time, however, it was to Germany that Europeans with serious scientific ambitions went for their higher training. German electrical industries and chemical industries alike established research laboratories, sometimes with governmental funding, and by the end of the century Germany's scientific-technological capabilities had surpassed those of Britain, a feat recognized most strikingly in the awarding of one-third of all Nobel prizes in physics and chemistry to German scientists before the Nazis' seizure of power at the beginning of 1933.[20]

What research laboratories helped to realize was the transition of science from its earlier identification with natural philosophy to its modern, contemporary identity. Unlike the variously trained, part-time students of nature who had made up the early generations of Royal Society membership and who, as wealthy gentlemen, had typically worked on their own as each saw fit, scientists in the nineteenth century sometimes worked as full-time professional specialists on research teams funded either privately or publicly. And as fields of knowledge and research grew, so too did the costs of training scientists and bringing them together in their work. What in the twentieth century would be called "Big Science" already was partly foreshadowed in some of its aspects a full century earlier, in the century David Knight calls "the age of science."[21] Research laboratories lay at the heart of this development, of course, but whereas we have thus far emphasized the role science played in the possibilities afforded technology and industry, we need now to switch the spotlight to glimpse how commercial production and technological advances benefited science in an increasingly specialized alliance.

Professional instrument makers certainly were not new in 1800. We have noted the important contribution of precision craftsmanship made by Jan Musschenbroek to the spread of Newtonian science. In this same vein, in the eighteenth century James Watt began as an instrument maker before becoming much more famously known for combining scientific theory and method with technological genius in his reconstruction and refinement of the steam engine. Yet the availability of precision instruments suffered serious supply problems even into the opening decades of the nineteenth century. It was then when the great experimental scientist Michael Faraday felt it necessary to alert fellow experimenters to the need for them to save up their materials for making their own further apparatus.[22] Scientific needs called forth commercial response, however, and in the course of the nineteenth century and thereafter the manufacture of precision instruments spread precision itself in the machined accuracy and uniformity of mass production. Machine manufacture supplied science with tools and apparatus furthering the investigation of nature in various fields where precision itself might count for anything knowable. Needless to say, industry served itself as well. Whether or not the experimentation ultimately led to anything of industrial use or commercial value, from the commercial standpoint the business of science was business.

What influence technology generally had upon science was in some instances no less profound than the impact new sciences like electromagnetism and chemistry had upon the creation of new technology or upon the industrial revolution as a whole. In some cases the arrival of an altogether novel apparatus opened the door to some emerging branch of science. One such case is interesting because it respects a piece of equipment that has long been a common fixture in the high school chemistry lab, namely, the bunsen burner. Prior to Robert Bunsen's invention of this particular tool, furnaces had been employed experimentally for heating purposes. The gross limitations of furnace use need hardly be pursued, but the point here in any event is that the clear flame of the bunsen burner allowed for the development of spectroscopy, which among other things connected the substantial investigation of the sun and the stars with the "flame testing" of substances here on earth.[23] No small matter, that. Yet probably the greatest impact technology had upon physical science in the nineteenth century involved no new technology whatsoever. Far from being new, the technology in this case stood at the very center of the industrial revolution, which is verily what made its theoretical implications so significant for science in general. This technology was the steam engine, along with various other devices connecting heat to motion in the mechanical operations of industrial production.[24]

The science arising from the study of technologies translating heat into work was thermodynamics, which soon became a fertile field for theoretical as well as applied science. Among the numerous scientists who became active in the field, Sadi Carnot is one of the better known for his formulation of ideas leading to the first law of thermodynamics, which established the principle of the conservation of energy. Thermodynamics brought with it, in turn, valuable knowledge pertaining to energy use in industry. Science historian John Gribbin summarizes the situation as follows: "Thermodynamics both grew out of the industrial revolution, which provided physicists with examples (such as the steam engine) of heat at work, inspiring them to investigate just what was going on in those machines, and fed back into the industrial revolution, as an improved scientific understanding of what was going on made it possible to design and build more efficient machines."[25] The case of thermodynamics well illustrates the interbreeding of science and technology made possible, of course, by the instrumental nature of maker's knowledge, which renders it useful in multiple contexts of application, be they experimental or industrial.

We have no interest here in cataloguing the various ways in which science and technology may have interacted while energizing the industrial revolution and moving it forward. Our discussion has wanted, rather, to indicate how each ally made major contributions to the other in the making of the modern world. Throughout the nineteenth century and beyond, such artifice sprang from the instrumental intersection of our knowledge of nature and the art of invention. If the French Revolution had spotlighted the artificial nature of human institutions in the political action of revolutionaries who unmade one society while seeking to make another in its place, the industrial revolution simply went about the business of remaking the material conditions of life in the West without any overall plan guiding it. This "dual revolution," as historian Eric Hobsbawm has so felicitously phrased these political and industrial phenomena, carried the West irrevocably along its modern course,[26] and from the standpoint of our interests in these chapters, the industrial revolution clearly bears the major responsibility for making our world what it has become. Under the reign of the *vita activa* the project of mastery unfolded without any master design, but with a very definite instrumental nature. Our relations with nature became increasingly instrumentalized while the institutional mediations shaping both our interactions with one another and the rhythms of our life-world became increasingly mechanized. Instrumentalization and mechanization went hand in hand, or, as Lewis Mumford specifies such developments: "Power, speed, motion, standardization, mass production, quantification, regimentation, precision, uniformity, astronomical regularity, control, *above all control*—these became the passwords of modern society in the new Western style [italics added]."[27]

This syndrome of integral traits identifies what Mumford elsewhere calls the "neotechnical" phase of Western civilization, which he sees emerging with the nineteenth century's "modern society in the new Western style."[28] Control is the axis around which turns the instrumental core of our modern society. In its making and operations this industrialized world has largely done without the institutional brain center we find in Bacon's utopia. Given the historically undirected nature of the modern project of mastery, no group of scientists or technologists has masterminded our development from any command center like that enjoyed by Bacon's House of Solomon. Such an authority on the face of it seems, in fact, anathema to the value of intellectual freedom embraced dearly by scientific communities ever since Galileo found himself embattled with the declining authority of the Roman church. In the tradition of modern science there has been no Rome nor anything resembling a House of Solomon that might have governed assent to scientific truth or commanded obedience to scientific method. No such institutional authority has been necessary largely because scientific authority properly obtains without institutional housing. Actual corruptions in hiring/firing and peer review aside, freedom of inquiry abides in the scientific tradition inasmuch as pragmatic standards of instrumental evaluation are meant by their very nature to provide authority.

At the center of the success and authority of Newtonian science beat the heart of the master, and just as industrialization throughout the nineteenth century spread progress through technological control, so too did Newtonian science advance knowledge through experimental mastery across numerous fields of study. Yet astounding irony exists at the experimental level of the development of physical science in the nineteenth and early twentieth centuries. On that level the very success of Newtonian physics drove scientific knowledge into realms of nature that would overthrow the Newtonian understanding of reality. Einstein's work was fully subversive on that score, and turning now to briefly consider that work and its impact, we will need first to get our bearings by recalling Newtonian science, not for purposes of review, but to see things that have yet to appear.

Probing the In-Between of Things

Right at the outset Newton's *Principia* drew criticism from philosophical quarters on the Continent, especially those occupied by rationalists. The ablest of these critics was G. W. Leibniz, whose thinking owed something to Descartes and who set his philosophical sails against the wind of Newtonianism before that wind became prevailing. Patrimony respecting the differential calculus divided Newton and Leibniz, for each had created the calculus independent of the other and viewed the other's mathematical offspring suspiciously, to say the least. Debate also arose, however, over substantive questions of how reality is put together and how it operates. Samuel Clarke, Newton's scientific disciple and theological champion, represented Newton in argumentative correspondence with Leibniz that carried over the course of ten letters that were published together in 1717.[29]

One issue concerned the nature of space and time. Newton embraced absolute space and absolute time as the metaphysical framework within which all mundane motion occurs in measurable, spatio-temporal magnitudes. This embrace was religious as well as philosophical since Newton viewed absolute space and absolute time as attributes of God.[30] Leibniz criticized Newton's rendering of God, and he philosophically attacked the idea of absolute space. As the century wore on, however, theological criticism subsided while the absolutes in themselves were retained as metaphysical presuppositions of understanding that simply had to be true. Looking backward from the twentieth century, Max Jammer puts the point this way:

> One thing is certain: Newton's mechanics, as expounded in the *Principia*, is one great vindication of his theory of absolute space and absolute motion.... The outstanding success of Newtonian mechanics in the physics and astronomy of the last two centuries seemed an indubitable guarantee of the soundness of its philosophical implications. It is not surprising, therefore, that the criticisms leveled ... against the theory of absolute space found no echo in this long period.[31]

Another issue pertained to certain celestial irregularities in Newton's universe that, left to themselves, would ultimately eventuate disorder and physical breakdown. God's intervention kept things aright as far as Newton's thinking went, but for Leibniz an imperfect Creation was unthinkable. What Newton had thought to be an imperfect mechanism proved, however, to be a problem of imperfect knowledge, as work by Laplace and Lagrange later smoothed out the mechanical wrinkles in Newton's universe, making it a mathematically perfect machine. In this particular case, the successful advance of Newtonian science is what made Newton's God superfluous; Laplace's having no need of the hypothesis of God assumed a universe that worked perfectly well on its own lawful causality. One issue that ultimately would not go away, however, concerned the attraction between or among material bodies. Whether the attraction concerned a thing's constituent particles holding together as a mass or the relations between the planets and the sun, how did such attraction work? Newton had thrust his law of gravitation into the in-between of celestial things, but how did gravity itself work?

Leibniz charged that, since the Newtonian laws required causality to operate through physical contact and since Newton offered no such explanation of gravity, his understanding of the in-between of things had to entail non–material action of some sort: it was action at a distance, either by divine miracle or some other occult force.[32] Without dismissing the possibility of divine action, Newton himself denied the physical possibility of action at a distance; and although he claimed that he would not "feign hypotheses" for want of an experimental handle on the issue,[33] he nevertheless in at least two instances went on to pose the possibility of there being some extremely rarified material medium — perhaps there is an ether — through which such mechanical action might be translated.[34] Newton simply did not know how gravity worked. But no matter, at least for the time. As the editor of a recent edition of the *Clarke-Leibniz Correspondence* has remarked, "By the end of the century these disputes had died away.... Newtonian science had succeeded in explaining the tides, the paths of the comets, and the irregularities in the movement of the moon; and therefore its theoretical foundation had to be secure."[35] Newtonian science appears to have silenced its early critics by leaving their issues up in the air. Science simply advanced. Instrumental knowledge needed only pragmatic authorization and predictive success to lay its claim to truth, and the very real progress of Newtonian science solidified such a claim. Or so it all seemed.

As reality itself appeared to Newtonian eyes, from the largest star to the tiniest atom, the universe is "thingish" through and through in the basic respect of its materiality: things = bodies = material masses, which operate dynamically through an impact physics entailing spatio-temporal magnitudes of position/distance and velocity/momentum/acceleration. This is the picture of things— indeed, the absolutely complete picture of things— Laplace's seer possesses in snapshot form, and it is interesting to note here that Laplace's superhuman Newtonian scientist, who knows everything there is to know worth knowing, pays no attention whatsoever to the question of gravity's cause, or to the larger question of action at a distance. In any case, Newtonians entering the nineteenth century stood on firm ground as they surveyed nature's massive terrain. That terrain might stretch far beyond what anyone could see, and exploration would be challenging, but Newtonians were confident explorers. Newtonian laws provided a sure compass for understanding new fields. Things would prove otherwise, however, for those nineteenth-century scientists who probed the inner workings of magnetism, electricity, and light, and whose work would ultimately throw physical science back upon some of the very issues it thought it had left behind.

Scientists working in those fields investigated phenomena lying in the in-between of things, and it is upon their shoulders that Einstein later stood as he pushed scientific understanding beyond the world of things. As Einstein would recall the development of nineteenth-century physics in an autobiographical reflection:

> Dogmatic rigidity prevailed in matters of principles: In the beginning ... God created Newton's laws of motion together with the necessary masses and forces.... [But later the] factor which finally succeeded, after long hesitation, to bring the physicists slowly around to give up the faith in the possibility that all of physics could be founded upon Newton's mechanics, was the electrodynamics of Faraday and Maxwell. For this theory and its confirmation by Hertz's experiments showed that there are electromagnetic phenomena which by their very nature are detached from every ponderable matter — namely the waves in empty space which consist of electromagnetic "fields."[36]

Einstein's work culminated century-long investigations of light, electricity, and magnetism, and followed upon the experimental insights of Michael Faraday and the mathematical genius of James Clerk Maxwell. The key to that development lies in the idea of "waves." Waves are not "things" in the Newtonian sense of particulate masses, and they fill apparently empty space in the manner of electromagnetic fields of force that, in turn, do not work as an impact mechanics. Pursuing the reality of waves ultimately would mean discovering that in the in-between of things, no "things" exist. But this understanding took time to crystallize in the work of these later giants.

At the turn of the nineteenth century Thomas Young revived the idea that light is a wave phenomenon rather than a particulate phenomenon, and around that same time experimental work by Coulomb and Oersted causally interconnected the phenomena of electricity and magnetism.[37] Oersted's findings did not seem to conform to Newtonian physics, but the French mathematician and physicist André-Marie Ampère, who invented the electromagnet and coined the term *electrodynamics*, managed to explain Oersted's findings in good Newtonian terms. Less conforming was the English physicist and chemist Michael Faraday, whose experimental work in electricity and magnetism during the second quarter of the nineteenth century led him to introduce what Einstein was later to call a "bold idea." Faraday abandoned Newtonian mechanics and spoke simply of magnetic lines of force filling empty space. Faraday, not being adept mathematically, had the further effrontery of presenting his ideas without the elaborate mathematical dress required for suitable science.[38] Mathematical physicists at this time were used to resolving electrical and magnetic currents "into point-masses in conservative motions serving Newton's laws," as Gillispie tells us, and looked upon Faraday's work "with either indulgence or a touch of scorn."[39] At mid-century the German physicist Hermann von Helmholtz spoke for the entire age when he proclaimed that scientific explanation found satisfaction only when phenomena were reduced to Newtonian laws.[40] The *Principia* remained biblical. Yet not for much longer.

Faraday's magnetic lines of force did find full mathematical formulation in James Clerk Maxwell's work during the century's third quarter. Maxwell integrated Faraday's findings in an elaborate set of mathematical equations satisfying all of the experimental data concerning the action of light, electricity, and magnetism. Bringing together the wave phenomenon of light and the magnetic lines of force led Maxwell to speak of this mathematical integration as the theory of the *electromagnetic field*. Maxwell's theory identified the in-between of things as an electromagnetic field in which transverse energy waves operate without any material medium. The reality of "empty space" identified by Maxwell's equations did without any ether. This conception of reality is essentially what would overthrow Newtonian understanding, though not all at once, nor even in Maxwell's own understanding. Maxwell himself, in fact, felt some real need to render his work acceptable in basic Newtonian terms, and so he superimposed a mechanical model on his field theory. Not committing to his speculative model, however, Maxwell distinguished its illustrative role (this *may* be real) from the explanatory nature of his field equations (these *are* real).[41]

Maxwell's equations of the electromagnetic field found direct experimental confirmation in Heinrich Hertz's discovery of the radio wave in the 1880s. Hertz had no need of any mechanical overlay, and his experimental discovery would help to convince some others of what Einstein later would declare quite decisively: "In Maxwell's theory, there are no material actors."[42] And if, like Faraday's lines of force, Maxwell's theory was able to do without mechanical explanation, then fields of electromagnetic energy waves might stand in the place of action at a distance without requiring any material medium for causality to operate. An experimental showdown with the ether therefore seemed warranted, leading to the famous Michelson-Morley experiment, which refers to what were perhaps three separate tests performed towards the close of the nineteenth century. Michelson-Morely's interferometer testified against the existence of an ether, which then led to efforts by the Irish mathematician George Fitzgerald and the Dutch physicist Hendrik Lorentz to save the ether via one or another formulation of physical relativity that, as Max Jammer puts it, "ascribes to the ether or absolute space certain definite effects which by their very assumed existence preclude any possible observation of the ether."[43]

By the turn of the twentieth century Newtonian understanding had reached the point where scientific reason was being taxed to find satisfaction. Modern science generally abides by the principle of Occam's razor, which refers to the preference reason has for explanations that are sharp and economical, as opposed to those that are overwrought, cumbrous, or burdened in any way by *ad hoc* non–essentials. Copernicus had held this advantage over Ptolemy, and now, centuries later, Occam's razor was being dulled. Maxwell's mechanical overlay was unnecessary to the experimental findings accounted for in his field equations, and the Michelson-Morley findings were being interpreted to subvert their testimony against an ethereal medium. In the historical context of the early twentieth century, following Occam's razor would mean cutting scientific understanding away from Newton's rational mechanics. That great achievement was Albert Einstein's.

Einstein's thinking took shape at the turn of the twentieth century. He remembered that time with some modesty when he later reflected:

> One got used to operating with these [electromagnetic] fields as independent substances without finding it necessary to give one's self an account of their mechanical structure; thus mechanics as the basis of physics was being abandoned, almost unnoticeably, because its adaptability to the facts presented itself finally as hopeless.... [But e]nough of this! Newton, forgive me; you found the only way which, in your age, was just about possible for a man of highest thought—and creative power. The concepts, which you created, are even today still guiding our thinking in physics, although we now know that they will have to be replaced by others farther removed from the sphere of immediate experience, if we aim at a profounder understanding of relationships.[44]

Einstein's reference to the sphere of immediate experience is telling, for it points to the distance the physical investigation of nature had come since Newton's time and that of the eighteenth century. Science had arisen among generations of investigators who had handled things quite directly in their experimentation, and, unsurprisingly, their mechanical explanation of the motion of bodies in space and time conformed immediately to the human experience of having to lift and move things. Later scientists investigating the in-between of things still handled things, but these things increasingly had become the instruments needed to measure, determine, and predict what they themselves could neither see nor immediately manipulate. This instrumentalization of observational conditions was very much a part of the nineteenth century's alliance of science and technology, and is what permitted scientific understanding to reach what Einstein called a "profounder understanding of relationships."

Einstein's theory situates objectivity within observational contexts that relativize appearances.[45] Einstein's famous thought experiments, or *Gedenkenexperimenten,* illustrate this point: imagine, Einstein says, this one event seen first from this coordinate system, or spatio-temporal point of observation, and then from this other coordinate system, and understand how that physical event obtains different spatio-temporal appearances relative to those perspectives. In the General Theory of Relativity the physics of the continuous field conditions the relativity of all empirical space-time measurements.[46] The field concept itself reconstitutes the foundation of physical reality. In Einstein's words: "A new concept appears in physics, the most important invention since Newton's time: the field. It needed great scientific imagination to realize that it is not the charges nor the particles but the field in the space between the charges and the particles which is essential for the description of physical phenomena [italics deleted]."[47] Instead of trying to imagine how the in-between of things might be understood in terms of things themselves, as good Newtonians did, Einstein explains both things and the empty space between them in terms of a continuous electromagnetic field of energy: "By far the greatest part of energy in concentrated in matter; but the field surrounding the particle also represents energy, though in an incomparably smaller quantity"; hence, he continues, "Matter is where the concentration of energy is great, field where the concentration of energy is small."[48] Energy is the basic reality of the continuous field and of all things physical.

Newtonians had always known that inertial mass and gravitational mass were equivalent, but it was Einstein's genius to realize the significance of that equivalence for explaining accelerated motion in terms of energy.[49] This fundamental conceptual move enabled Einstein fundamentally to reduce Newtonian reality to the terms of electromagnetic energy, and even though Einstein himself admitted the need for dualism in addressing *field* and *matter*, he sought to generalize the theory of the gravitational field in a mathematical structure that would satisfy the total field.[50] Einstein never succeeded, but his work gained people's allegiance nonetheless. Relativity theory entered a culture already mature scientifically, and, within that context, it immediately impressed people on a number of predictive fronts where "crucial tests" decisively displayed the superiority of relativity physics over Newtonian understanding.[51] Moreover, relativity theory could account for both the success of Newtonian science as well as the problems bedeviling that science. Relativity addresses phenomena possessing or approximating the speed of light, and, at the same time, it underwrites the validity of the Newtonian laws for the study of things possessing much lesser velocity. Newtonian mechanics thus could become a "special case" within relativity theory's profounder understanding of relationships.

Since science is pragmatic at its core, one way to interpret this relationship between relativity science and Newtonian science is to think of science as a "self-correcting" tradition. This reading can be misleading, however, when we consider the question of truth, or, more precisely, the truth-status of relativity theory. And truth surely is the issue for science once Einstein jettisons Newtonian understanding in favor of relativity thinking. Truth is not an issue of relativity itself; nothing could be more wrongheaded than the loose idea that Einstein "relativized truth." Relativity theory concerns appearances, not truth, and had Newton himself formulated relativity physics three centuries ago, relativity itself would have constituted truth on Earth. The problem Einstein's work posed for truth, rather, is this: in secular revelation scientific authority had delivered truth, and once scientific authority later came to overturn that selfsame truth, it lost its claim to truth just as it lost its claim to the certainty of its explanation of physical reality. Just as Einstein told Newton that he had been wrong, so there may come along at some later date some scientific explanation that repeats the same judgment about relativity theory.

Relativity theory may seem "true" because it has proved so productive and become a habit of thought, but so much — nay, more!—could be said for two centuries of Newtonian science as well. Indeed, we will not know the truth to be the truth even if we possess it because science has before proved its own understanding wrong. A newer theory may be better predictively, but such improvement does not *correct* explanation in any sense of making it certifiably right or true.

This loss of truth in scientific understanding runs throughout the science philosophy of our time. Most revealing is the ubiquitous use of the word *theory*, a term Einstein himself used to identify the status of his own beliefs. The belief that scientific understanding has grasped nature's reality purchases no secure quarter in the idea of *theory*. Theory recognizes *interpretation* in its *models*, or *constructs*. Telling words these, for they raise the issue of *meaning* for instrumental understanding. For example, while Newtonian science and relativity theory are both pragmatically successful, the very meaning of space and time differs radically in the two different theoretical frameworks. Where this problem appears amidst the rarified debates within science philosophy, much of which entails language philosophy and symbolic logic, two central issues arise: one issue concerns the relationship between theoretical terms and observational terms (i.e., interpretation and description of physical events), and the other concerns the relationship between differing explanatory constructs (i.e., how can Newtonian science and relativity physics relate conceptually?).[52] In this last regard philosophers of science argue over the development of scientific knowledge (historical reconstruction), over the nature of inter-theoretic criticism between competing conceptual constructs, and over the possibility of there truly being "decisive experiments." However recondite such matters may be, what they manifest is simple to understand: scientific reason is taxing itself to philosophical extremes to make sense of what scientific understanding entails in the absence of truth.

Most influential among all twentieth-century philosophers of science are the two thinkers who faced the dissolution of truth head on: Karl Popper and Thomas Kuhn. Popper thought out his perspective on science in the 1920s and 1930s in opposition to the Vienna Circle of Logical Positivists. Articulated in the 1920s, logical positivism sought to unify all knowledge, that of the humanities as well as that of the natural sciences, upon the truth-criterion that until then had prevailed in physical science.[53] A pragmatic principle of verifiability was to ground the meaningfulness of all statements claiming knowledge of any kind: if a statement cannot be verified in some pragmatic way, it is utterly meaningless and inadmissible as knowledge. Popper's thinking was directly influenced by Einstein's destruction of Newtonian truth, and the striking irony of the positivistic identification of meaningfulness with verifiability was not lost on him.[54] For Popper, the testability of a knowledge claim or of a theoretical explanation could do without truth. An experimental test cannot verify—it cannot establish truth—but it may falsify; it may disprove. Using this insight into the meaning Einstein's work holds for the interpretation of science, Popper went on to elaborate his doctrine of falsificationism, which attempts a logical reconstruction of scientific discovery and theory formation on the basis of experiments intended to refute knowledge claims.

Popper's widespread influence soon was challenged directly by the work of Thomas Kuhn, whose *Structure of Scientific Revolutions* proved to be every bit as influential as Popper's own work.[55] Kuhn also starts from a cultural-historical appreciation of the impact of Einstein's physics, but he departs from Popper's "critical rationalism" in his reading of how science works. Whereas Popper sees formal-logical operations governing theory formation and the development of scientific knowledge, Kuhn sees socio-psychological factors working within a community of scientists.[56] Kuhn's early work spotlights the nature of human belief. Scientists abide by *paradigms* (very loosely, belief structures) that govern, and are reinforced by, the normal everyday work performed within a community of scientists. A revolutionary shift from an old to a new paradigm has less to do with crucial experiments than with a "gestalt switch" in the socio-psychological attachment people have to their beliefs. In Kuhn's work, then, the loss of truth enables us to see science as an institutional practice bearing the same admixture of rationality and irrationality normal to any human enterprise. Far from realizing secular revelation, science in the twentieth century comes to realize its fully human character.

Truth be gone, in light of Einstein's work scientific understanding becomes at best a working theoretical interpretation of reality within a community of scientists. In Popper's words, "all laws, all theories, remain essentially tentative, or conjectural, or hypothetical, even when we feel unable to doubt them any longer."[57] Einstein surely never doubted that the General Theory of Relativity was true, but it is hard to imagine any descendent of Alexander Pope's poetizing Einstein's *theoretical interpretation* as meaning secular revelation. Having his name become synonymous with genius would have to suffice in Einstein's case. The times had changed. But not everything significant in or about science had. More directly, the admitted theoretical status of understanding in relativity physics in no way alters the fact that relativity science continues to uphold the deterministic nature of both nature's operations and our knowledge of them.

Although Einstein's explanation of things walks on different theoretical legs than does Newton's, relativity physics and Newtonian science do, nevertheless, trod the same ground of pragmatic testability and are in complete agreement on the formal character of causality. Einstein keeps the "if ... then" schematic very much alive, maintaining the logical symmetry of explanation and prediction. F. S. C. Northrop catches the thrust of this point exactly when he writes, "[F]or Albert Einstein scientific method entails the validity of the principle of causality ... [in the sense that] with the empirical determination of the present state of a system, as defined by theoretical physics, the future state is logically implied."[58] Echoes of Laplace ring loudly here, and well they should since here we find the critical connection, the essential likeness, between Einstein's work and Newtonian science. For as long as the reality underlying appearances partakes of a strictly deterministic causality, then science, even without any claim to truth, continues to authorize mastery of nature. When all is said and done, that is, Einstein said no only to the meaning in Newtonian theory, not to its Baconian character.

Relativity physics may be fully in keeping with the "if … then" schematic upheld by Newtonian science, but the same cannot be said for the work of other early twentieth-century scientists who were investigating nature beyond the world of things. Whereas relativity thinking addresses physical reality principally at the macrophysical level, it is in the microphysical realm that scientists would find themselves operating outside of the "if … then" schematic. And in that netherworld, where microphysicists probe the nature of subatomic events, the reality being observed was essentially different from anything Einstein or his theory could admit. In fact, words like "events," "observation," and "phenomena" could not even apply as they ordinarily do in the scientific investigation of nature. This strange development would have everything to do with the same instrumentalization of observational conditions enabling scientific understanding to stretch beyond the world of things in the first place. The problem with microphysical stretching, however, is that somewhere along the instrumental route to the subatomic realm, men came to believe they had lost their very ability to understand nature at all.

The Tragic Insight of Quantum Theory

Seldom do revolutions formally announce themselves, but the radical upheaval in the scientific community caused by quantum physics did enjoy an official introduction.[59] Many of that community's brightest lights gathered in Brussels in 1927 to attend the Solvay Conference, which, under Hendrik Lorentz's leadership, had decided that year to spotlight subatomic physics. Among the leading microphysicists there, Niels Bohr and Werner Heisenberg very quickly took center stage. They had been working as part of a team of subatomic physicists under Bohr's direction in Copenhagen, and at the Solvay Conference Bohr unveiled what instantly became known as the "Copenhagen interpretation of quantum mechanics." The so-called Copenhagen interpretation would become our quantum theory in all essentials, but when it first appeared, Bohr's theory shocked the entire scientific community. It caused an earthquake, shaking that world to its foundations and causing fissures that divide the scientific world even today, almost a century later. No scientific event before or since would have such an immeasurable effect.

Quantum theory sprang from the experimental work Max Planck had earlier performed with black body radiation at the turn of the century.[60] Planck's work had addressed an extremely strange phenomenon concerning the intensity of radiation being emitted from an incandescent solid body. Spectrographic waves measuring such intensity showed discretely discontinuous magnitudes of energy. This phenomenon contradicted all normal and expected physical action in the same way a thermometer would be suspect were it to register a rising or falling temperature only at five-degree intervals—as if the warming of a winter morning could occur instantaneously, changing immediately from 0 degrees to 5 degrees centigrade without passing through the intervening values 1 through 4. Puzzling though the phenomenon was, Planck faced the matter on its own terms and in 1900 announced that energy is emitted and absorbed as discontinuous quanta. Energy valuations, or "movement," obtains in jumps or leaps, as it were, and Planck determined the experimental value to be given this discontinuity, or *quantum of action*. "Planck's constant," symbolized by h, measures the indivisibility of quantum action.

We have spoken of bold ideas, and Planck's surely was one. Planck's constant proved immediately eventful in microphysical experimentation. The first really major event came with Einstein's work on the photoelectric effect, which involved the emission of electrons from metal being bombarded by light. Explanations of the effect had made no headway until Einstein decided to determine the effect in terms of h, which proved directly successful. Success was costly, however, or at least problematic for the larger scientific understanding of light. Whereas everyone had come to understand light as a wave phenomenon, Einstein's experimental use of Planck's constant required light to be particulate, or thing-like. How could a wave, which is undulatory in nature, act like a thing, which is not undulatory? Einstein made no pretense of explaining this impossibility, calling his explanation heuristic and allowing for both descriptions in the scientific investigation of light. Heuristic it certainly was, for soon the dualism of particulate and undulatory descriptions pervaded microphysical science. More particularly, widespread experimentation with the electron showed a fundamentally glaring inconsistency, in fact a blatant contradiction. The electron was schizophrenic in its experimental action, behaving on some occasions like a wave and at other times like a thing. Such schizophrenia suggested, of course, that no one actually knew what the electron was, or how to explain it. The electron could not really be crazy; that is, the reality of the electron certainly could not be both a wave and a particle. Or could it? And if it could, then in what sense would it make any sense to say that it could?

Answering these questions and related ones was the essential task Bohr and his associates had undertaken in Copenhagen, and the answer they came upon was the quantum theory he and Heisenberg unveiled to the scientific community at Solvay. It was not an answer pleasing to that community, nor was it then or at any point after a theory easy to understand even in its broad outline. This displeasure and our difficulty in generally understanding the quantum theory in everyday terms spring alike from the paradoxical aim lying at its very center: the theory intends scientifically to explain microphysical reality in terms of science's experimental inability to explain that reality. Here we have a paradox badly in need of clarification, clarification we might begin by extending our metaphor: quantum theory originally addressed the schizophrenia of subatomic nature by way of microphysics' own experimental self-analysis. Using this idea as our guide, we now need to follow Bohr and Heisenberg along the terribly slippery path of quantum theory's experimental self-analysis.

Heisenberg recalls the opening of his most fundamental insight into quantum physics as follows: "We had always said so glibly that the path of the electron in the cloud chamber could be observed. But perhaps what we really observed was something much less. Perhaps we merely saw a series of discrete and ill-defined spots through which the electron had passed. In fact, all we do see in the cloud chamber are individual water droplets...."[61] Heisenberg's description of what microphysicists experimentally observe is entirely literal, and this literal standpoint provides the clearing in which Heisenberg and Bohr erected the structure of quantum theory. Microphysical science reaches into the subatomic realm of nature by means of sophisticated instruments that serve as the hands and eyes of experimental practice. The conditions of observation are thoroughly instrumentalized, such that all the scientist ever does see are instrumental registrations of one sort or another. Quantum theory recognizes this fact and then elaborates its implication for the meaning of *experimental observation*. The instrumentalization of observational conditions throws the meaning of experimental observation back upon itself like a reflecting mirror. Experimental observation no longer refers to any direct relationship between the scientist and the natural phenomenon he or she is investigating; rather, it refers to the relationship between the observer and the instrumentation through which the scientist encounters nature in his or her work. And what a difference this self-referential character of quantum theory makes for the very meaning of words!

The self-reflexive character of experimental observation in quantum theory makes our understanding of quantum theory especially difficult right at the ground level of communication, for when Bohr and Heisenberg speak about observation or phenomena, they mean by these terms something altogether different from what is terminologically intended and ordinarily understood in scientific discussion. The problem, in fact, is even more radical than this point might only loosely suggest. As Bohr tells us in retrospect, "I warned especially against phrases often found in the physical literature, such as 'disturbing of phenomena by observation' ... since words like 'phenomena' and 'observation,' just as 'attributes' and 'measurements,' are used [by us] in a way hardly compatible with ordinary language and practical definition," and as he continues, "As a more appropriate way of expression I advocated the application of the word *phenomenon* exclusively to refer to the observations obtained under specified circumstances, including an account of the whole experimental arrangement."[62]

Phenomenon refers to the experimental connection between microphysicist and nature under specified conditions of instrumental *observation*. In Heisenberg's words, that is, when all is said and done, "Whenever we try to deduce laws from our study of atomic phenomena, we discover that we no longer correlate objective processes existing in space and time, but only observational situations."[63] The question "but what about nature's reality?" is impermissible from the standpoint of what can be known because, to speak more generally, quantum knowledge refers to some kind of *experimental reality* as it were. This experimental reality, this integral connection between the subatomic realm and the conditions of experimentation, underlies the self-analysis quantum theory undertakes when it addresses the schizophrenia of the electron. Bohr's "principle of complementarity" wants to explain that schizophrenia by holding that the *nature* of the electron is both wave-like and thing-like. The *nature* of the electron is a function of the particular experimental arrangement employed to probe the electron. Given one specified experimental arrangement, the electron will register an undulatory action; given another kind of experimental apparatus, it will register particulate action. Complementarity thus asserts that our knowledge encompasses mutually exclusive *descriptions* that in themselves reflect mutually exclusive experimental realities.[64] Bohr's analysis does not, of course, render the electron understandable in itself. In Erwin Schrödinger's imagery, "however we think of it, it is wrong; not perhaps quite as meaningless as a *triangular circle*, but much more so than a *winged lion*.'"[65] According to quantum theory, the electron certainly is unthinkable in itself, which is just what Schrödinger wants us to understand. Yet Schrödinger's own thinking can mislead us here inasmuch as quantum knowledge severely restricts our description to experimentation and thus prohibits our speaking of nature as some *Ding an sich*.[66] Theory offers no therapy in Bohr's analysis of the electron's schizophrenia.

The electron, in fact, is even crazier than Bohr's analysis might seem to suggest at first, for whether its action be wave-like or thing-like, the electron displays the character first noted by Planck, jumping around unpredictably in quantum leaps. The men from Copenhagen addressed this problem no differently, employing the same experimental self-analysis, but with eyes now concentrated on the problem of causality. And here the crux of the experimental reality of microphysics finds focus in the term *interaction*. The experimental reality is constituted interactionally. Our only contact with microphysical phenomena "down there," considered in themselves momentarily for the purpose of discussion, is by means of our instruments, and, from the standpoint of our experimental knowledge, the instruments themselves in some unspecifiable way interactionally contribute to the *reality* of those *phenomena.*

Quantum theory holds that the very act of measuring partly determines the outcome of the experiment and, further, that this interactional event necessarily entails some inaccuracy. As Heisenberg tells us, "We could for instance be interested in the motion of an electron through a cloud chamber and could determine by some kind of observation the initial position and velocity of the electron. But this determination will not be accurate" since, inevitably, "we cannot completely objectify the result of an observation.... [W]hat happens depends on our way of observing it or on the fact that we observe it."[67] Magnitudes of space and time fall subject to complementarity because our means of determining precisely the position of an electron renders its velocity indeterminate, while, conversely, any exact specification of velocity necessitates our affecting the electron's position in some indeterminate way. How this interaction might be affecting things lies, of course, beyond any possible instrumental determination because any such determination would suffer the same problem of instrumental contact: in other words, no more than humans are instruments able to jump over their own shadows. Heisenberg addressed this large problem in his "uncertainty principle," which is also called the "principle of indeterminacy."

At the most general level, Heisenberg's principle of uncertainty addresses the status of microphysical knowledge. Whatever we can say or know about the subatomic realm is inescapably uncertain, as this uncertainty is interactionally grounded in our instrumental encounter with microphysical nature. This generalization then reduces to operational meaning in the specific mathematical applications the principle has within the experimental reality of subatomic physics. Heisenberg himself operationalized this interactional uncertainty in a probability calculus, using Planck's constant (h) as the irreducible value of interactional indeterminacy. As Heisenberg tells us:

> A probability function is written down which represents the experimental situation at the time of the measurement [and this] ... represents a mixture of two things, partly a fact and partly our knowledge of a fact.... It should be emphasized, however, that the probability function does not in itself represent a course of events in the course of time. It represents a tendency for events and our knowledge of events. The probability function can be connected with reality only if one essential condition is fulfilled: if a new measurement is made to determine a certain property of the system.... Therefore, the theoretical interpretation of an experiment requires three distinct steps: (1) the translation of the initial experimental situation into a probability function; (2) the following up of this function in the course of time; [and] (3) the statement of a new measurement to be made of the system, the result of which can then be calculated from the probability function.[68]

Heisenberg here spells out the self-reflexivity of the experimental reality, and he does so directly in terms of how the probability function governs microphysical measurement. The experimental "given" is a fact, but what this experimental situation in itself brings about—what the act of measuring (inter-) actualizes as an experimental reality—is nothing ascertainable in the given. Such indeterminacy requires an experimental follow-up in order to determine which one of the possible experimental realities obtaining within the probability calculus was, in fact, (inter-)actualized. That experimental reality is not understood to be what takes place "down there" in between experimental determinations. As Heisenberg would remind us of the meaning of a microphysical *phenomenon*, "we have to realize that the word 'happens' can apply only to the observation, not to the state of affairs in between observations."[69] The *causality* of what occurs lies within the indeterminate interaction of the experimental reality, which means that probability is one with the act of measurement and not an effect of incomplete knowledge (as if we might obtain more information for a complete description of state). What this situation means, in turn, is the ultimate paradox of quantum theory articulated in Bohr's principle of complementarity. Bohr writes:

> On the one hand, the definition of the state of a physical system, as ordinarily understood, claims the elimination of all external disturbances. But in that case ... any observation will be impossible, and, above all, the concepts of space and time lose their immediate sense. On the other hand, if in order to make observation possible we permit certain interactions with suitable agencies of measurement, not belonging to the system, an unambiguous definition of the state of the system is naturally no longer possible, and there can be no question of causality in the ordinary sense of the word. The very nature of the quantum theory thus forces us to regard the space-time coordination and the claim of causality, the union of which characterizes the classical theories [Newton's and Einstein's], as complementary but exclusive features of the description.[70]

Einstein's world is gone, just as is Laplace's. There is no causality belonging to microphysical nature as we can objectively know it, nor is there a space-time reality belonging to microphysical nature as we can objectively know it. According to the men of Copenhagen, microphysical nature unto itself simply has no objective reality of which we can speak. Adhering strictly to such understanding is problematic at best, hence Bohr's concern to italicize or put in quotation marks terms like "phenomenon" and "observation." More generally, however, speaking about the electron or of microphysical nature leads all too easily, if not inescapably, to bad faith. Bohr and Heisenberg themselves often forgot the experimentally grounded point of *as we can know it* and spoke of the indeterminate subatomic realm as if its experimental reality were other than experimental. Interesting in this regard are the different frames of reference implied by the uncertainty principle and the principle of indeterminacy: the former respects the state of our knowledge whereas the latter principle tends to point towards something "down there."[71] These different references are easy to conflate since it is all too human to speak of microphysical nature as something "down there," all by itself, with or without our interactional encounter with it.

Perhaps the most illuminating example of this very human need to speak of that "reality down there" is Whitehead's metaphorical rendering of the electron and its physical "action." "It is as though an automobile, moving at the average rate of thirty miles an hour along a road, did not traverse the road continuously; but appeared successively at the successive milestones, remaining for two minutes at each milestone."[72] This metaphor works brilliantly in illuminating our inability to make sense of physical action outside of normal spatiotemporal coordinates. However, it still assumes some realistic ground for imagining the electron outside an experimental reality. The root of the problem lies, of course, in the unfathomable nature of the interaction itself. Could our instruments speak and tell us of their encounters with nature, quantum theory might mutate or perhaps disappear instantly in the face of reports on reality. Despite all of their power and fine registrations, however, our instruments are dumb. Quantum theory thus finds its speech in everyday language, but brings meaning to that speech out of utter darkness.

What the men of Copenhagen arrived at in their experimental self-analysis is something akin to the "tragic insight" of which Nietzsche spoke where he faintly glimpsed science's reaching some limiting point at which scientific knowledge would recoil upon itself and bite its own tail.[73] Quantum theory comes to something like such insight as it ponders its own enterprise in the absence of an experimentally determinable, objective world "out there." What quantum theory realizes in its experimental self-analysis is a limitation that nature has imposed on scientific investigation at the subatomic level. Whatever on earth subatomic nature in itself may or may not be, it limits our experimental capacity to disclose an ascertainably objective reality. We encounter ourselves in some measure in the interactional nature of that experimental reality: we bite our own tail, and unavoidably so. This tragic insight of quantum theory has been the deepest source of division within science from the very moment Bohr and Heisenberg formally delivered their theoretical child into the hands of the scientific community at the Solvay Conference.

A most unwanted child it was then, and even as an octogenarian, it today still finds itself unwelcome in some quarters. Right from the start resistance to quantum theory not surprisingly focused on the issue of *understanding*, which immediately concerns the reality and determinability of a causal, spatiotemporal reality — an *objective* reality — in microphysical nature. The nature and parameters of this issue were established instantly at Solvay in the arguments between Bohr and Einstein. Heisenberg recalls that debate in a lengthy reflection, as follows:

> We all stayed at the same hotel, and the keenest arguments took place, not at the conference hall but during the hotel meals. Bohr and Einstein were in the thick of it all. Einstein was quite unwilling to accept the fundamentally statistical character of the new quantum theory. Needless to say, he had no objections against probability statements whenever a particular system was not known in every last detail.... How-

ever, Einstein would not admit that it was impossible, even in principle, to discover all the partial facts needed for the complete description of a physical process. "God does not throw dice" was a phrase we often heard from his lips during these discussions. The discussion usually started at breakfast, with Einstein serving us ... another imaginary experiment by which he thought he had ... refuted the uncertainty principle. We would at once examine his fresh offering, and on the way to the conference hall ... we would clarify some of the points and discuss their relevance. Then, in the course of the day, we would have further discussions of the matter and, as a rule, by suppertime we would have reached the point where Niels Bohr could prove to Einstein that even his latest experiment failed to shake the uncertainty principle. Einstein would look a bit worried, but by next morning he was ready with a new imaginary experiment more complicated than the last, and this time, so he avowed, bound to invalidate the uncertainty principle. This attempt would fare no better by evening, and after the same game had been continued for a few days, Einstein's friend Paul Ehrenfest ... said "Einstein, I am ashamed of you; you are arguing against the new quantum theory just as your opponents argue about relativity theory." But even this friendly admonition went unheard ... Einstein had devoted his life to probing into that objective world of physical processes which runs its course in space and time, independent of us, according to firm laws.... And now it was being asserted that, on the atomic scale, this objective world of time and space did not even exist and that the mathematical symbols of theoretical physics referred to possibilities rather than to facts.... Later in his life, also, when quantum theory had long since become an integral part of modern physics, Einstein was unable to change his attitude—at best, he was prepared to accept the existence of quantum theory as a temporary expedient. "God does not throw dice" was his unshakable principle, one that he would not allow anybody to challenge.[74]

The question of whether or not God plays dice may be of significance in the private recesses of faith, but that question meant absolutely nothing in the argument joining Bohr and Einstein as scientists. The issue was, and ever since has been, whether or not science can devise an experiment that might grasp subatomic nature and hold it to account. Until his death decades later, Einstein steadfastly maintained that there must be such an experimental possibility, and over the years this selfsame position or argument has been held by legions of like-minded scientists who have refused to embrace quantum theory as anything but a temporary expedient. Quantum theory finds acceptance from this perspective only as a stopgap measure that "explains" all of our experimental findings, but only at the expense of understanding; when some experimental key for unlocking subatomic secrets arises, the jettisoning of quantum theory will be celebrated with great abandon. Against this position, Bohr simply asserted the tragic insight Einstein refused to accept: "The limit, which nature herself has thus imposed on us, of the possibility of speaking about phenomena objectively, finds its expression, as far as we can judge, just in the formulation of quantum physics."[75]

Einstein and Bohr spelled out the issue quite formally in 1935 in an exchange in the journal *Physical Review*. There, Einstein very precisely identified the conditions defining what it means to understand physical reality objectively. He wrote, "If, without in any way disturbing a system, we can predict with certainty (i.e., with probability equal to unity) the value of a physical quantity, then there exists an element of physical reality corresponding to this physical quantity."[76] This assertion rests completely upon the foundation of understanding underlying classical astrophysics: there is symmetry between explanation and prediction. Einstein's own relativity physics is fully in keeping with this symmetry. From the perspective of quantum theory, however, Einstein's position assumed the very knowledge denied microphysics by the experimental reality quantum theory articulates. As Bohr replied, "[T]he *finite interaction* between object and *measuring agencies* conditioned by the very existence of the quantum of action ... requires the necessity of a final renunciation of the classical ideal of causality and a radical revision in our attitude toward the problem of physical reality...."[77]

The issue found no resolution in this exchange, but the burden of proof lay entirely with Einstein. Indeed, the crux of the issue was less who was right about there being an objective reality than whether or not Einstein or anyone else could provide an experiment that might sidestep the difficulties deemed inescapable by the men of Copenhagen. On that score Bohr used the point of quantum theory to puncture each and every one of Einstein's thought experiments. Given the pragmatic standards of scientific authority, Einstein's critique of quantum theory failed because none of his tests would work, neither at Solvay nor over the remaining years of his life. At every experimental turn, quantum theory's tragic insight into the limitation of science turned aside Einstein's challenge. Yet Einstein persisted, and his persistence has always found explanation in the famous line "God does not throw dice." Einstein's faith in the "Old One," as he occasionally referred to the idea of God, was not, however, the real faith sustaining his lifelong argument with Bohr. His real faith sprang absolutely from the success of modern science and had to do altogether with his unquestionable belief in our experimental ability to disclose the objective reality behind what we see. Surely scientists are able to design an experiment that can grasp nature's microphysical causality! Without doubt nature's defiance cannot be unyielding! Meanwhile, nature has proved to be every bit as persistent as was Einstein. Indeed, on this particular issue nature's obduracy has withstood the efforts of many scientific assailants over these past eight decades.

Beyond the Means of Mastery

Decades ago the science philosopher Norwood Hanson remarked on the opposition to quantum theory as having created "a cemetery of dead *Gedenken-experimenten*."[78] The number of graves has multiplied greatly since Hanson's observation, and a walk through that cemetery would be long indeed. Upon the headstones of the failed experiments or experimental proposals appear many illustrious names: Einstein's, Popper's, and those of a host of first-rate physicists. Popper at one point even charged the men of Copenhagen with not having tried sufficiently to refute their own theory; they had not practiced enough falsificationism! Paul Feyerabend, another science philosopher, rightly pointed out that quantum theory had been articulated fully on the basis of countless failed experiments that had been carried out in the best tradition of good scientific testing.[79] Popper's point was in bad faith anyway, for the obvious fact is that neither he nor the legions of like-minded naysayers were able to succeed where quantum theory failed. As quantum theory might predict, they were bound to fail, for as Bohr tells us, "According to the quantum theory, just the impossibility of neglecting the interaction with the agency of measurement means that *every observation introduces a new and uncontrollable element*" [italics added].[80] Experimentation thus inherently falls short of the critics' demand for success because the limitation of our experimentation is immediate with the fact that the conditions of observation are themselves not entirely controllable.

Einstein and those following his charge against the quantum theory quite understandably connected the possibility of explaining physical reality to the need for experimental disclosure and certain prediction. Their perspective kept full faith with the scientific tradition upholding the symmetry of explanation and prediction. But the central issue of control—the question of whether or not nature possesses any independence from experimental command—never received its due even while lying at the very surface of the tumultuous debates capsizing physical explanation. The point of nature's independence never became decisive, surrounded as it was by the turbulence caused by quantum theory's announcement that in its instrumental plunge to the unfathomable depths of nature, modern science had forever lost its capacity to understand physical reality as a whole. The question of there being an objective reality or the issue of our understanding that reality is what captured everyone's attention. Our own interest here, however, is with the loss of control over the conditions of observation in microphysical science, that loss being identical with our inability to control the question of causality. In this very regard Heisenberg's uncertainty principle formally expresses the absence of an experimental command over the realm of appearance.

Nature directly manifests its defiance by denying microphysics command over the effect of experimentation. J. Robert Oppenheimer puts the matter pointedly when he tells us that "there is no complete causal determination of the future on the basis of available knowledge of the present. The application of the laws of quantum theory restricts, but does not in general define the outcome of an experiment."[81] The "if ... then" schematic of Newtonian physics and relativity theory thus disappears in the breakdown of strict predictability. No control over the question of causality exists within the experimental reality precisely because, as we have already discussed, within that reality, causality entails some indeterminate interaction between nature's stuff and scientists' probing measurements. Years ago in his book *The Revolution in Physics*, Louis de Broglie in this same regard summarized the quantum picture as if he were directly criticizing the presumption of the Laplacean seer, as follows:

> Knowing with precision the values X_O, Y_O ... of the quantities characterizing a system at an instant T_O, we could [in classical physics] predict without ambiguity what values X,Y ... would be found if they were determined at a later instant T. This resulted from the basic equations of the physical and mechanical theories and from the mathematical properties of these equations. This possibility ... implying that the future was in some manner contained in the present and adds nothing to it, constituted what has been called the determination of natural phenomena. But ... it is this knowledge which quantum physics considers impossible.... Having determined the values of the quantities characterizing a system at an instant T_O ... the physicist cannot predict exactly what the value of these quantities will be at a later instant; he can only state what the probability is that a determination of the quantities at a later instant T will furnish certain values.[82]

Heisenberg's principle of uncertainty expresses the probabilistic nature of our causal knowledge. This uncertainty is identical to the experimental situation where a knowledge claim can find substantiation only once the experimental reality is followed up later in time in order to determine what that experimental reality effected. This point stood stage center in Heisenberg's presentation of microphysical experimentation in three acts. As we recall Heisenberg's words in this critical regard, an experiment requires "(1) the translation of the initial experimental situation into a probability function; (2) *the following up of this function in the course of time*; [and] (3) the statement of a new measurement to be made of the system, the result of which can then be calculated from the probability function [italics added]."[83] Because knowledge of the future is not synchronous with knowledge of the present, the necessary staging of experimental knowledge itself takes place in time. And what this inclusion of time in experimentation means quite literally is the temporalization of the conditions of observation. Unlike the experimental knowledge of the master, which stands without time inasmuch as it enables the master to know exactly what is to follow by causal command, knowledge binding the terms of observation to temporal conditions in and of itself evinces the absence of mastery. Quite simply, time becomes telling where mastery goes wanting.

Various kinds of experimentation exist, and the Baconian version (cum mathematical objectification) has monopolized our attention until now for the very obvious reason that it underwrote the rise and success of modern science. The experimental manipulation and control of nature's causality crowns the achievement of modern science and remains the ideal model for the scientific acquisition and use of instrumental knowledge. But quantum physics could not stay the course of maker's knowledge in experimentation; subatomic nature would not allow that. With the dissolution of the "if ... then" formula of maker's knowledge, a different kind of knowledge appears at the foundation of physical science, one expressing an asymmetry between explanation and prediction. "A final kind of discomfort with Copenhagen concerns an asymmetry in microphysical explanation and prediction.... *After* a microphysical event *X* has occurred within our purview, we can give a complete explanation of its occurrence within the total quantum theory. But it is impossible to predict the features of *X* so easily explained *ex post facto*."[84] Norwood Hanson is not saying that explanation becomes sensible in terms of any meaningful *ex post facto* explanation of what occurs within the experimental reality. Rather, his point is that the effect produced becomes determinable as an experimental reality within the calculus of probability, but only *after the fact*.

Explanation may be a risky term to use in the context of quantum theory in the same way words like *phenomenon* or *observation* can lead the unwary widely astray. In any event, the point here is that such knowledge is post–dictive rather than pre–dictive. Knowing what occurs comes only after seeing the effect, and certain knowledge of this kind is partly a child of time; it reports what is past. Post-dictive knowledge hardly has the commanding presence of hard and true predictive knowledge, and certainly it is not meant for everyone's liking — not for those, anyway, who like their physics deterministic, orthodox and Newtonian, as Hanson puts it.[85] But such knowledge is not chosen, so to speak. It is made necessary, fitting, and appropriate wherever nature refuses to conform to the instrumental command of scientific experimentation. Opposition to the men of Copenhagen is understandable both psychologically and historically up to some point of reasonable belief, but beyond that point such opposition betrays the attitude of the master. That nature should have no say in the matter; that nature simply must yield, if not today, then tomorrow or the day after — such commanding disregard for nature expresses the very belief and attitude that should have been overturned by the quantum revolution. In that revolution of the early twentieth century, the experimental claim to mastery came to an end in our inability to control the question of causality or command the realm of appearance. Nature quite plainly said no to the will of the would-be master.

During the two centuries unfolding between Newton's death and the Solvay Conference modern science and industrialization gave Western peoples every reason to believe that our species will have its way with nature and make this world over as it sees fit. Nothing has substantially overturned this deeply embedded cultural impression. Nature's announcing its independence from instrumental command almost a full century ago has meant little or nothing amidst the triumphant thickening of human artifice and the overwhelmingly impressive wizardries of maker's knowledge on all research fronts. Quantum theory itself has made possible impressive wide-ranging applications of instrumental knowledge; without it, there would be no Google.[86] Indeed, our instrumental recipes now take up volumes that would make the *Encyclopedia* appear but a brief notation on human possibilities. But what irony we have here. For after Solvay, modern science could no longer authorize the project of mastery.

We need to be very clear about the point here. Belief in mastery culturally arose upon the ground of applied science when those promulgating Newtonian science spread the practical word. More fundamentally, however, the authorization modern science provided such belief rested in principle upon nature's yielding its causality in experimental contexts wherein science might command the realm of appearance. Whatever the practical yield of quantum theory has been does not concern us, because, from the standpoint of this study, the significance of the quantum revolution lies in nature's proclaiming its independence from human command. Nature formally announced for the first time that it possesses causality escaping instrumental control. And as if nature knew that this point needed repetition, it soon would offer countless examples of its independence in biological and ecological causalities eluding our instrumental grasp. We will examine a number of these offerings in chapters to follow. Here, however, we need to note that the quantum revolution fathered another first, namely, belief in domination as pure faith.

Over and against nature's stunning proclamation of its independence we witness the first real expression of belief in mastery as a pure faith — that is, belief in science's virtually unlimited capabilities vis-à-vis nature's causalities became faithful precisely at the point when scientific experience no longer legitimated such belief. Einstein's refusal to accept what quantum theory has to say amounted to a dismissal of the claim nature was making to its own independence. Science would make nature yield, if not immediately, then down the road. And down the road legions would follow Einstein's lead, with many contributing further to that cemetery of dead *Gedenkenexperimenten* of which Hanson spoke. An interesting cemetery it is, for in it should stand in prominence a headstone for the passing of the scientific authority that had underwritten our belief in mastery. But no such headstone can be found. In fact, nowhere can be found even a death notice, much less an obituary. Belief in mastery has survived regardless of scientific experience; this Newtonian legacy has lived on as illusion. Seeing this belief at work in our current relationship to nature, and exposing it for the illusion it is, is the critical task of the remainder of our study.

4

Science in the Land of Sometimes and Perhaps

A parallel may seem to hold between our own experience and that of the early moderns who lived in that twilight time before the advent of modern science. We, like they, no longer possess truth; and our time, like theirs, finds uncertainty at the bottom of things. But this parallel lacks substance beneath the surface. The essential difference separating our time of experience from theirs is deep: while the birth of modern science elevated Western belief above uncertainty and appeared to deliver truth, with the early twentieth-century developments in physics we witness science itself disavowing certifiable truth claims and delivering us our uncertainty. Yet, this essential difference notwithstanding, just as our forefathers found science liberating, so we too should find science emancipatory, at least to the extent that its very historical development frees us from the errant beliefs of earlier scientific understanding. And, indeed, there can be no doubt that we have been freed from a great deal that was wrong or misleading in Newtonian science. When it comes to mastery of nature, however, we still generally remain bound to the belief that science authorizes human dominion. At this point such belief amounts to real illusion.

Belief in mastery persists in the conventional terms materialized by modern industrialization. Science and its applications subjugate nature to human employments, and the project of mastery advances without plan or supervision. In this view of things nature is simply *there* for human exploitation, and, given time and support, science can handle virtually any question or problem nature may pose. Persistent as this belief may be, it takes a backseat today to formulations of mastery far bolder in presumption and promise. Today there are two different programs expressly offering scientific command centers to govern nature as a whole. It is immaterial whether or not the control panel at either command center is fully developed; in either case, it is a design on nature aiming to put this planet under self-conscious human control. Illusion becomes dangerous at this point, and we will critically tackle these proposals of mastery in later chapters. Before sallying forth in that direction, however, we need first to acquaint ourselves with a new understanding of science arising in the middle third of the twentieth century upon the ruins of Newtonian science. This new understanding of our instrumental relationship to nature is widespread today within both the scientific community and Western culture as a whole. It is also the understanding providing meaning for those programs currently proposing human dominion.

The present chapter will explore the origin of this new understanding of science in those same early-twentieth-century developments in physics we investigated in the preceding chapter. Having already witnessed the collapse of Newtonian science and discussed the loss of truth claims in the wake of Einstein's work, we will turn to consider what primary lesson the scientific community has drawn from the revolution wrought in Copenhagen. While that lesson is not exactly or primarily the one we ourselves drew from the quantum revolution, it will help us to reorient our discussion in general. More directly, during the middle third of the twentieth century the scientific community generally formulated an understanding of science in the shadow of quantum physics and proceeded at once to clothe that understanding in biological meaning. In this chapter our discussion will at once come upon new scientific terrain in its encounter with biology, but this is the very terrain for which we shall want some compass since in coming chapters we will pursue the belief in mastery onto the turf of living things. Preparing such a compass also will be the work of the present chapter, making this chapter a Janus-like passageway between the story of science just ended and the investigation of organic change yet to come.

Instrumentalism in the Name of Science

Relativity physics turned truth into theoretical interpretation and certain understanding into predictive theory. Yet relativity gave the scientific community reason to celebrate the advance of science beyond the world of things. Theoretical interpretation might fall far short of the secular revelation that Newton had delivered, but Einstein's theory of relativity accounted for everything Newton had explained in mechanical terms and so much more. Science gave away the claim to authorize truth in exchange for greater theoretical depth and predictive scope, and this advance by relativity physics was a major gain for science as a whole. In this same vein, even reflective minds are usually subject to habitual belief, and "theory" has a way of assuming the feel of "truth" when, through time, its successes lead to its becoming a paradigm of normal science, to employ Kuhnian terminology. In any case, relativity physics bore understanding beyond the world of things, and even though Einstein never achieved a unified field theory and some current thinking suggests that relativity theory may be flawed, no one would deny the tremendous expansion of knowledge made possible by Einstein's work. In the case of quantum mechanics, however, matters of gain and loss, or trade-off, seem more difficult to sort out in the large historical picture.

Einstein and those who would follow his lead contra Copenhagen saw in the admitted unintelligibility of quantum theory's experimental reality, the very unintelligibility of science itself. Put differently, they saw a revolution in sensibility that undermined what science had always assumed in its investigation of nature, namely, a reality to be discovered to greater or lesser extent in the work of science. What can be made of science if there is no objective reality "out there" of which we can speak? Einstein's relativity theory was mathematical through and through, but at no point did it ever leave any doubt about there being natural elements that are the objective correlates to its equations. Quantum theory, on the other hand, was about rendering the experimental situations themselves in statistical terms and leaving out of account any serious concern for either the reality of nature or the nature of reality outside of those situations. As Brian Davies writes of quantum theory's formalism, "We make no conjectures about what fundamental particles themselves are really like, but content ourselves with knowing how accurate various mathematical models are in different circumstances."[1]

When science finds itself subject to philosophical reflection at this level of meaning, the debate about quantum theory and science as a whole has been between the "realists" and the "instrumentalists," this latter term being given to those whose thinking generally abides by what quantum mechanics maintains. Along with those issues already mentioned in connection with Popper and Kuhn in the previous chapter, the difference between realism and instrumentalism defined the central focus of much science philosophy in the second half of the twentieth century. From the standpoint of instrumentalism, science is defined by the nature and function of scientific theories, and the *sine qua non* of scientific theories is that they allow us to predict what our laboratory instruments and apparatus will register in an experimental situation.[2] Instrumentalism in this basic sense quite clearly articulates quantum theory's self-analysis as a perspective on science itself, and, not surprisingly, Bohr and Heisenberg were among its earliest major proponents. Whatever interpretive content a theory may possess bears no significance whatsoever except as a predictive function, be that strictly deterministic or not. The essential concern of theory is the (re)construction of experimental arrangements under the rule of computational-predictive schemata. Applying instrumentalist thinking to concepts rather than theories gave rise to "operationism," which simply operates at a lower level of generalization where concept-formation is thought to entail the production of performance rules.[3] Operationism might be either subsumed under instrumentalism or viewed as a variant of this more general perspective. Over and against instrumentalism, realism adheres to the more traditional idea that science is above all about something entirely separable from science and knowable as such. Realism does not dispute the interpretive character of theory; it assumes, rather, that scientifically meaningful interpretation concerns some core aspect(s) of nature's reality. Writing on the philosophy of science, J. J. C. Smart puts the matter this way (where Vo denotes observational vocabulary and V_t denotes theoretical vocabulary):

> The chief realist objection to instrumentalism is ... that on the instrumentalist view it is a surprising and inexplicable fact that the world is such that *Vo* facts can be related to V_o facts by means of V_t sentences, whereas on the realist view there is nothing mysterious at all: if the V_t sentences really are about the hidden mechanisms of the universe after all, then *of course* they can be expected to help us to relate V_o sentences to other V_o sentences.[4]

The other major objection to instrumentalism comes from Popper. Popper is a realist whose particular criticism of instrumentalism comes from his critical rationalism. Popper's critique is direct: "If theories are mere instruments of prediction we need not discard any particular theory even though we believe that no consistent physical interpretation of its formalism exists," and, as he continues, "instrumentalism is unable to account for the importance to pure science of testing severely even the most remote implications of its theories, since it is unable to account for the pure scientists' interest in truth and falsity."[5] All realists' objections take most seriously the core reduction verily given with the word *instrument*alism: from the perspective of instrumentalism, however mathematically elaborate and consistently visualizable a scientific theory may be, it is essentially an instrument, a tool. Critics of instrumentalism thus intend to save science from a technological identity, both to uphold pure science as an investigation of reality and to make explicable the history of modern science, considered as something quite other than a history of technology.

The Baconian-manipulative nature of modern science stands fully exposed in the instrumentalist version of science: maker's knowledge is intimately connected to the instrumental handling of nature, and with the increasing instrumentalization of observational conditions throughout the past two centuries, the very doing of science has become more and more technological in its nature and appearance. This fact does not gainsay the failure of the instrumentalist version of science to explain why science arose in the first place or developed as it did. Pragmatic though its instrumental knowledge may be, modern science found articulation and development in interests that were not simply technological nor altogether reducible to technical capability. We should remember that Bacon himself held those experiments bringing light in higher regard than those bearing fruit. What must be highlighted here generally, however, is that the instrumental nature of science stands forth dramatically when reality and truth are abandoned and the reason for doing pure science widely appears to become problematic. Problematic to be sure, but hardly out of the question in actual fact. Many scientists still do pure science.

Human curiosity knows no limit, nor should it. Science still promotes among many the wonder and truth-seeking that maintains the distinction between science and technology. Today "string theory" has a large number of proponents who believe that this idea may possibly unite everything in heaven and earth in a few grand mathematical equations. The fact of there being no apparent way of directly testing this idea experimentally seems to be of no great consequence to some string theorists. Quantum cosmologists carry on in various directions, some of them arguing for the coexistence of many worlds or parallel universes.[6] Uncertainty and the experimental conundra of quantum physics are not eliminated in such pursuits, but given interpretations involving multiple dimensions having only mathematical formulation. In a none-too-admiring fashion former *Scientific American* writer John Horgan refers categorically to all such science as "ironic science" and to those doing such science as "ironic scientists." He writes of the more naïve of these that they "possess an exceptionally strong faith in their scientific speculations, in spite of the fact that those speculations cannot be empirically verified. They believe they do not invent their theories so much as they discover them; these theories exist independently of any cultural or historical context and of any particular efforts to find them," and then later, "a few diehards ... will practice physics in a nonempirical, ironic mode, plumbing the magical realm of superstrings and other esoterica and fretting about the meaning of quantum mechanics. The conferences of these ironic physicists, whose disputes cannot be experimentally resolved, will become more and more like those of that bastion of literary criticism, the Modern Language Association."[7] Anyone even remotely familiar with the "lit crit" of the late twentieth century will sense at once the level of Horgan's disdain.

Pure theory is not to be denied even when we consider the subatomic realm, which essentially gave birth to instrumentalism in the first place. Yet instrumentalism will not be denied either. In point of fact, instrumentalism generally has carried the day. One science philosopher remarked some time ago that by the mid–twentieth century instrumentalism had already become the "in-group view,"[8] and to hear Edmund Bolles tell it, once quantum physics had wrought its revolution, "physicists put away all pretense of striding closer to knowing what was out there beyond ourselves."[9] Among physicists and science philosophers alike, then, instrumentalism is alive and widespread as a matter of fact, if not as a satisfactory theoretical or historical perspective on the question of why giants called natural philosophers once walked along trails leading to modern science, or of why anyone would want to know why nature either generally or in some particular respect is as it is, and not otherwise.

While Bolles' statement is far too sweeping and somewhat misleading, the point is that in the wake of quantum theory large numbers of practicing scientists identified their practice of science with the belief that theory, whatever its content, obtains essential value only insofar as it enables us to predict, certainly or probabilistically, the effect of our experimental handling of some research question. Science basically becomes problem-solving in this regard, and at one level or another we are all instrumentalists whenever we speak metaphorically of our "conceptual tools" or "theoretical tools." Needless to say, when speaking in such a fashion we are certainly a long way removed from the sensibility that obtained in secular revelation. David Lindley summarizes the situation as follows:

> As has been true since the 1920's, questions of interpretation and philosophy simply do not arise for the great silent majority of physicists who apply quantum mechanics to their endeavors. In the late nineteenth century, especially among scientists educated in the German tradition, there was a feeling that as theoretical physics advanced, it ought to evolve a philosophy along with it. Nowadays most physicists are reared in the Anglo-Saxon style, steer clear of Plato and Kant, and are belligerently uninterested in what philosophers make of their theories.[10]

While twentieth-century physical science played the chief role in establishing the instrumental perspective among practicing scientists—who are, after all, generally neither philosophers nor historians of science—during the second half of the twentieth century, increasing numbers of scientists found themselves and their careers directly subject to "Big Science," which turned science and scientific knowledge into tools on a grand public stage. Big Science refers to an alliance of science and technology, but it is not the same institutional alliance we glimpsed earlier when discussing the development of nineteenth-century science in the context of the industrial revolution. That earlier alliance has grown immensely, feeding directly into the research and development divisions vital to modern industry. Big Science, on the other hand, came about largely as a wartime product of an alliance between science and government, and much of the publicity science received during the second half of the twentieth century came from the service that science was giving to national security, as a tool of national defense. Whereas the alliance of science and industry has historically had a relatively low profile, in Big Science the instrumental utility of science has been pronounced in headlines the world over. Big Science quite simply broadcast for social consumption an instrumental understanding of science during the same decades that such an understanding was taking hold, albeit on different terms, within the scientific community.

Big Science conveys its meaning terminologically, as it regards extremely large research projects requiring immense funding for sizable teams of scientists who utilize huge and/or hugely expensive apparatus. "The high cost of scientific instruments, facilities, and payrolls made Big Science affordable only to government agencies or international consortia, drawing influence away from the universities, societies, and philanthropies that had been the main supporters of scientific research before World War II."[11] But opinions differ on the origin of Big Science. Gino Segrè sees Big Science emerging directly from the womb of physics and locates its birth precisely in 1932 with the experimental use of the first cyclotron, which involved large numbers of men, money, and material.[12] Others more commonly look to Los Alamos, New Mexico, identifying the introduction of Big Science with the Manhattan Project and the development of the atomic bomb. Alvin Weinberg certainly had the United States' Department of Defense on his mind when he coined the term Big Science in a seminal piece appearing in the July 21, 1961, issue of *Science*, where he complained of Big Science's undermining the institutional and intellectual independence of traditional science programs and practices by way of "journalitis," "moneyitis," and "administratitis."[13]

Wherever we spot its debut, a great deal of the Big Science carried out in the United States during the second half of the twentieth century sustained the so-called military-industrial complex, which connected both private industry and university work to military purposes within a widely expansive network of defense projects and government funding. Big Science grew rapidly as the engine of the Cold War's spiraling arms race, and once President Kennedy identified space as the "new frontier" just three years after the Soviets had launched Sputnik, Big Science became pioneering beyond the Cold War that Kennedy himself was projecting into space. Kennedy's new frontier created a huge space industry, launching all sorts of projects in the wide-ranging research benefiting from the large financial fallout. Given such space largesse, Big Science radiated outward, increasingly subordinating scientific research to non–scientific aims in broad public view. Headlines the world over showed science to be a major instrument of interests paying little or no attention to science apart from its service to national defense, its technological utility, and its profitability.

Vitalized by war or threat of war for over four decades, this new marriage of science and technology suffered no annulment when the Cold War ended in the early 1990s. Needless to say, the Cold War had not been a war to end all wars, nor did the disintegration of the Soviet Union cause the Pentagon to close down. But in the early 1990s Big Science was beginning to set off in a new direction anyway, as serious interest in biological research arose. Defense still very much mattered, of course, and defense thinking quickly developed novel projects along biological lines. For example, the Defense Department's "Defense Advanced Research Projects Agency," or DARPA, which reported an annual budget of three billion dollars in 2005, currently undertakes widespread research in "bio-revolution" projects. DARPA now has an Unconventional Pathogen Countermeasure program, a Brain-Machine Interface program, and a Metabolic Engineering program, to name but a few.[14] People in DARPA speak generally about their work in "human enhancement," and regardless of its contextual irony here, this term and the interests lying behind it readily expand beyond the confines of defense needs.

The United States federal government boldly announced its interest in human enhancement when it pulled the financial plug on the development of the extremely costly super collider in 1993, only to turn around a short time later to redirect hundreds of millions of dollars to the Human Genome Project, which set out to map the entire human genome. Military uses of genetic knowledge ultimately may prove to be every bit as significant to defense as have been the contributions of high energy physics over these past six decades, but both up front and substantively, the aim of the Genome Project had little or nothing to do with defense while promising all kinds of human benefit. Indeed, the Department of Energy and the National Institutes of Health, which helped pay for the Project, were joined by molecular biologists, journalists, and science writers from far and wide in talk of cures for cancer and other dreaded diseases, of medical advancement on all fronts, and of the general possibilities of genetic intervention to enhance our lives either directly or indirectly in countless ways.[15] The Project itself actually broke the mold of Big Science because much of the molecular biology and computer research that went into the genetic mapping fell outside the official program and proceeded without any large centralized authority or governmental oversight.[16] In this respect the Project itself was an augury of the way of the world in genetic research, as molecular biology can connect with human enhancement along all sorts of private financial pathways leading to the burgeoning biotech industry.

Perhaps the pursuit of human enhancement outside of Big Science will offset to some degree the concern Weinberg voiced in his original article where he warned that "cultures which have devoted too much of their talent to monuments which had nothing to do with the real issues of human well-being have usually fallen upon bad days.... [W]e must not allow ourselves, by short-sighted seeking after fragile monuments of Big Science, to be diverted from our real purpose, which is the enriching and broadening of human life."[17] On the other hand, there lies in the pursuit of genetic enhancement enrichment of a different sort, which has everything to do with the employment of science for ends other than scientific knowledge. Indeed, one might argue that the profit opportunities within the old military-industrial complex ultimately will pale in comparison to those now joining molecular biology to the financial hip of the biotech industries. The ever forthright population geneticist Richard Lewontin spotlights this point clearly where he discusses recent developments in biology as follows:

> It has been clear since the first discoveries in molecular biology that "genetic engineering," the creation to order of genetically altered organisms, has an immense possibility for producing private profit.... As a consequence of these possibilities, molecular biologists have become entrepreneurs. Many have founded biotechnology firms funded by venture capitalists. Some have become very rich when a successful public offering of their stock has made them suddenly the holders of a lot of valuable

> paper. Others find themselves with large blocks of stock in international pharmaceutical companies who have bought out the biologist's mom-and-pop enterprise and acquired their expertise in the bargain. *No prominent molecular biologist of my acquaintance is without a financial stake in the biotechnology business.* As a result, serious conflicts of interest have emerged in universities and in government service. In some cases graduate students working under entrepreneurial professors are restricted in their scientific interchanges, in case they may give away potential trade secrets. Research biologists have attempted, sometimes with success, to get special dispensations of space and other resources from their universities in exchange for a piece of the action. Biotechnology joins basketball as an important source of educational cash [italics added].[18]

What the emerging generations of scientists will have in mind when they think about scientific learning and professional specialization may well link career enhancement and profit enhancement more intimately than ever before was imaginable. Profitability does not solely have to define what science comes to mean under the career circumstances discussed here, though increasingly it well may. Steven Shapin's most recent work aims to dispel the academic bias against those scientists who are acting as entrepreneurs in profit-making enterprises with venture capitalists. Where Shapin speaks of the scientists' vision, passion, and commitment to changing the world for the better as being among the criteria moving venture capitalists to invest, he argues that "the vocabulary of virtue does indeed map onto that of monetary reward."[19] However one wants to judge this mixture of motives, the reality to which Lewontin and Shapin both point from their different perspectives is one no different from that fostered by Big Science: scientific knowledge is understood as an instrument to be employed, in every sense of that term, for purposes lying largely outside the pursuit of science itself. So much was true, of course, of the industrial alliance between science and technology that got fully underway in the early industrial revolution, but in the case of Big Science and the biotech world, society's use of science as an instrument has had a public presence, a headlined reality, far greater than any before. It is in these biotech industries as well as in the continuing projects of Big Science that the instrumentalist understanding arising within the scientific community finds its primary institutional analogues in our society today.

Pure science and/or knowledge for the sake of knowledge will stay alive for as long as human curiosity remains healthy. Whatever may become of (super)string theory, some science always will transcend what many call "technoscience." Yet instrumentalism, nevertheless, has made a very deep impression, and instrumentalism's failure to account for why a Descartes, a Newton, or an Einstein pursued knowledge bears little significance in our contemporary culture. In fact, the critique of instrumentalism in its own way has been quite out of touch with scientific experience in recent time. For once we entered the second half of the twentieth century, if not before, the real thrust of Popper's criticism of instrumentalism already had become largely historical, attenuated to the point of nostalgia by our society's widening use of science as a means, as a tool, for achieving things other than knowledge of nature.

Naturalizing the Meaning of Instrumentalism

At the very time when instrumentalism was becoming the "in group" understanding of what science fundamentally is about, the mathematician Jacob Bronowski published a work of just under 150 pages that titularly bound our understanding of science to common sense. This short book, *The Common Sense of Science,* was published in 1950 and helped launch Bronowski's long career as a highly successful popularizer of science. *The Common Sense of Science* expressly sought to blunt the painful feeling of loss accompanying the dissolution of truth and reality in physics. Indeed, Bronowski turned that loss into a positive judgment on the development of science in the twentieth century: recent science has liberated us from the Laplacean illusion concerning reality and our knowledge of it, and now, in the clear light of common sense, we can see science in the larger context of human life. What makes Bronowski's work especially useful to us is that he articulates his judgment on the very terms where we ended our discussion in the previous chapter. Bronowski is worth following at some length, then, as he helps us further unfold the instrumental meaning science has obtained in the wake of twentieth-century physics.

Bronowski pulled no punches as he pondered the scientific landscape at mid-century. He dramatically observed that "[we] have been swept by a great wave of pessimism, which arises from our own feeling of helplessness in the recognition that none of us understands the great workings of the world. As science and knowledge have been broken into pieces, there has come upon us all a loss of nerve."[20] This loss of nerve was a direct consequence of twentieth-century scientific developments' having destroyed the very psychological and cultural comfort science itself had earlier provided. Bronowski addresses this discomfort metaphorically when he tells us, "We seem to be in a land of sometimes and perhaps, and we had hoped to go on living with always and with certainty."[21] Reminiscent these words may be of the time giving birth to modern science when Donne lamented, Pascal trembled, and Descartes sallied forth. No quest for certainty was on Bronowski's mind, however, for the uncertainty of our time is an experience with science already in it; our microphysical knowledge proclaims uncertainty and endorses it. And this fact is the very axis around which Bronowski seeks to turn the meaning of science 180 degrees—away from loss, towards profit; towards liberation from errant belief and a real understanding of science. Having believed for so long in a world that simply does not exist, we are now free of Laplacean illusion.

> The principle of uncertainty ... shook us all a good deal. After all, it said that nature could not be described as a rigid mechanism of causes and effects. And I recall again

> that all the successes of science, Newton's success and those of the nineteenth century, seemed to have been won hitherto by fitting nature with just this kind of machine. To say suddenly that at bottom the causal chains are not true, that the whole thing cannot be done—that seemed a strange discovery and a disagreeable one.... There simply is no sense in asserting what would happen if we knew the present completely. We do not and plainly we never can. This is precisely what the principle of uncertainty says to modern physics.... [Ultimately w]hat we have really seen happen is the breakdown of the plain model of a world outside ourselves where we simply look on and observe.... When that point was reached in astronomy, Einstein's laws took the place of Newton's.... And this is just what the principle of uncertainty showed in atomic physics: that event and observer are not separable.... [T]he world cannot be isolated from itself: the uncertainty *is* the world. The future does not already exist; it can only be predicted.[22]

With science standing at the center of unpredictability and prediction in this context meaning something less than certain knowledge, Bronowski proceeds to show how science, now that it has dispelled Laplacean misunderstanding and reconnected humankind to the world, can throw light on our relationship to the world and provide us with precisely the kind of knowledge required for success in our worldly dealings. Science in some sense becomes appropriate treatment for the psychological discomfort it has caused. Once again we need to listen to Bronowski at some length.

> Science is experiment, that is orderly and reasoned activity. The essence of experiment and all of science is, that it is active.... This of course is not peculiar to science. All living is action, and human living is thoughtful action. If this is plain enough as a statement about living, it still needs to be understood about science: that science is a characteristic activity of human life....This seems to me the most important point which I can make.... The characteristic of living things is that their actions are pointed towards the future....All this is hidden in the process of life; but it becomes plain and explicit when we look for scientific laws ... [because] a scientific law differs from our habitual way of pointing our actions toward the future only in being more systematic and explicit.[23]

Here Bronowski articulates the life context in which the instrumental identity of science—or, in a word, instrumentalism—finds a natural home, so to speak. Considered a predictive tool, science is the rationally honed expression of our need to interact with the world around us in purposeful, yet uncertain, action. "What we are looking for, in science as much as in the day-to-day of our lives, is a system of prediction: is, as it were, a predictor.... [I]t is perfectly possible to base a system of prediction on no principle except trying to get the right answer. This is exactly what all plants and animals do."[24] Prediction for Bronowski obviously means something less than certain knowledge, and in this discussion of his work we follow his meaning, which assumes that living things possess some kind of "predictive" capacity. The predictive tools in which science consists may distinguish and elevate the character of our purposeful action relative to the behavior of other species of life, but only insofar as our anticipation of what lies ahead finds self-conscious determination.

At this point in his reflection, in the chapter bearing the same title as the entire work, Bronowski identifies science with a manner of learning by trial and error. All practical learning consists in trial and error in the same measure that the future is uncertain and can only be predicted. Such learning is true of a living species as a whole, of individuals, and of science alike; and where Bronowski draws together these three domains of learning, he thrusts his point all the way home by naturalizing the meaning of science. "There is in all this a bold analogy between the way in which individuals learn, *the way in which species adapt themselves*, and *the way in which science works*. But, of course, it is my point that this is not merely an analogy: it is a true and close relation [italics added]."[25] In short, Bronowski articulates the idea of *instrumental adaptation*, to employ our own phrasing for what this understanding of science means. In the meaning of instrumental adaptation, science is our species-specific way of dealing with an interactionally changing and uncertain world in which we are temporally enmeshed and to which we must continually adapt.

We have spent considerable time listening to Bronowski because his reflections were among the first to identify generally what science has come to mean in the West since the middle of the last century. Moreover, the terms on which Bronowski unfolds this meaning conform very neatly, though not quite fully, with the analysis obtaining in the preceding chapter. The salient features of that conformity are worth recapitulating here before attending to anything else: Moving beyond the world of things with increasingly instrumentalized conditions of observation, science reported in the twentieth century that we cannot speak of a completely knowable physical reality; that human beings and nature are bound together interactionally; that the future is not given with the present, which perforce means our living with uncertainty in a world where our knowledge is temporally contingent; and that science fits us into this world with instrumental capabilities having greater or lesser predictive value. Bronowski articulates these lessons of twentieth-century physics and provides them with the naturalistic meaning we have in mind in the idea of instrumental adaptation.

The idea of instrumental adaptation cannot account any better than instrumentalism for either the history of science or for scientists' doing pure theory simply for the sake of knowledge. This fact should not surprise, however, inasmuch as the idea of instrumental adaptation immediately assumes an instrumental understanding of science and naturalizes it outside any cultural context of interpretation. In other words, when instrumental adaptation is viewed outside the interpretive-analytical narrative of our present work, it is nature, not any cultural-historical account of modern Western experience, that informs the idea and gives it meaning. This point comes through clearly in the words of George Gaylord Simpson, one of the West's leading mid–twentieth-century paleontologists, who had this to say of science at the same time Bronowski was undertaking his own reflections:

> [A]ll science is partly biological. It is all carried on by human beings, a species of animal. It is in fact a part of animal behavior, and an increasingly important part of the species-specific behavior of *Homo sapiens*. From the functional point of view, it means adapting to the environment. It is now, especially through its operating arm, technology, the principal means of biological adaptation for civilized man. It is an evolutionary specialization that arose from more primitive, prescientific means of cultural adaptation, which in turn had arisen from still more primitive, prehuman behavioral adaptation.[26]

Instrumental adaptation is what science has come to mean in Western culture for both scientists and the general public alike, and innumerable variations of the idea can be found in both professional and popular literature on the nature, meaning, or purpose of science. The 1968 edition of the *International Encyclopedia of the Social Sciences* had this to say under the entry "science":

> Because "science" is a term for an activity as well as ... a body of knowledge, we ought not to be surprised if it is continuous with the prescientific activities directed to the same ends, and if these are an extension of prelinguistic adaptive behavior.... What leads us, then, to introduce concepts at a given level? Not just the pressure of their existence on our blank brain. Rather, the preconscious selective effect of certain environmental and internal pressures begin the process of concept formation in our ancestors, as in other organisms....[27]

And as this same encyclopedia explains what science is or does thirty-three years later in its 2001 edition, "science produces *useful models* which allow us to make often useful *predictions*. Science attempts to *describe* what is, but avoids trying to *determine* what is (which is for practical reasons impossible). Science is a *useful tool*.... It is a growing body of understanding that allows us to contend more effectively with our surroundings and to better adapt and evolve as a social whole as well as independently."[28]

All sorts of variations on the theme of instrumental adaptation could be catalogued here, and we will have occasion to sample a fair number of them in what lies ahead. Our interest at this juncture lies, however, in a lacuna in Bronowski's work that is altogether glaring from the perspective of the present study. The lessons Bronowski draws from the historical development of modern science conform very nicely to our own analysis and interpretation, yet they fall short of what for us was the central lesson of the quantum revolution drawn in the preceding chapter. Quite simply, nature defied the experimental science that for over two centuries had authorized belief in mastery of nature. That defiance itself lies at the heart of the scientific developments bringing Bronowski in the first place to reflect on our having to live in a land of sometimes and perhaps. Yet nowhere does Bronowski spotlight nature's intransigence. If we wish to draw scientific lessons to liberate us from more than two centuries of errant belief, then surely not least among those lessons should be found some meaningful discussion of our having lost command over the realm of appearance; some discussion of our having met more than our match in nature's unyielding resistance to our interest in controlling its causality. Liberation ought to mean in principle that science withdraws its endorsement of our domination of nature and, respecting the authorization of human dominion, tells us like Poe's Raven, "Nevermore."

The very idea of instrumental adaptation ought to bring mastery of nature into question, at least from a terminological standpoint. Looking directly at the idea of *adaptation* alone should suggest that control is very much an issue. Adaptation and control are extremely different, if not totally antithetical, notions. We may adapt things we control to suit our own purposes, but ordinarily we say that we ourselves must adapt to something precisely because that particular something stands in some way or to some extent outside of our control. This meaning appears in the commonplace coupling of "adapt" with "have to," as in "you'll simply have to adapt," or "she will have to adapt to her new circumstances." If we were able to control the circumstances requiring our adaptation, we would not have to adapt ourselves in the first place. At the species level this selfsame meaning obtains: we understand that in order to survive, species "have to adapt" to changing conditions that they themselves may influence, but which they certainly do not control. Such meaning escapes reflection, however, where illusion prevails and the meaning of words and ideas falls prey to distortion. This distortion is just what occurs, in spades no less, when the idea of instrumental adaptation combines with instrumental programs proposing to control instrumental adaptation itself. But before attending to proposals identifying mastery of nature with instrumental adaptation, we need first to outfit our own critical understanding with the biological clothing suitable for discussing the study of living things.

The study of living things—or, more precisely, how one should study living things—is itself the very issue that today divides the house of scientific biology between those who follow the Newtonian tradition and, in so doing, open the meaning of instrumental adaptation to belief in mastery, and those who see in living phenomena and organic change something not entirely reducible to the maker's knowledge of modern science. We will approach the study of biological phenomena via this fundamental cleavage within biology and, in so doing, adapt the terms of our critical understanding to the biological and ecological domains we intend selectively to explore in the chapters ahead.

Approaching a Scientific Divide

Turning to the stuff of biology brings us to new territory. Shifting attention to the organic world does not leave behind the stuff of physics and chemistry. Chemistry is fundamental to biology while the physical world overall is vital to the phenomena of adaptation and evolution, and to the general study of ecology as a whole. In the study of living phenomena, however, a fundamental issue has cleaved the community of scientists for some time now. This issue is whether or not living nature entails elements of causality different from those typifying the causality found in inanimate nature. Is there, in other words, something in the nature of organic causality distinguishing it from the cause-effect relationships ordinarily defining the natural world investigated by classical physics? A second issue concerning scientific understanding then follows this question. This issue is whether or not biological study needs involve in addition to the conceptual perspective and experimental determinations of physics and chemistry, some mode of understanding and method of investigation different from that employed by physicists. In short, does biology fully conform to the form and nature of scientific understanding provided by modern astrophysics?

Belief that living things are in some defining sense *sui generis* is hardly new,[29] and in the nineteenth century Goethe wrote insurrectionary volumes on plants and animals intending to overthrow any and all Newtonian governance on the organic front. Two generations later Henri Bergson undertook a similar enterprise in his organismic idea of an *élan vital.* We will not venture there, nor in any direction requiring us to argue that living things possess some special animus or vitalism. The issue for us is causality, and our own discussion will stay within the confines of arguments carried on within the biological sciences. Indeed, causality arises as an issue at the very foundation of modern biology where the basic terms of a radical dispute over causality were established at the moment Darwin's understanding and Mendel's experimentation joined together to provide modern biology its initial scientific pedigree. Darwin's own fieldwork and the geological and taxonomical work on which Darwin's thinking drew combined narrative elements with rigorous empirical investigations. Darwin's evolutionary understanding, therefore, rooted itself outside the manipulative experimentation exemplified in physics and chemistry. Mendel's work in genetics, on the other hand, followed the experimental tradition of physical science. When the marriage of evolutionary understanding and genetic study took place, then, the possibility of divorce already loomed large in the very different methodological backgrounds of the parties wed. All that was needed to eventuate a split would be the smashing success of one party, whose career trajectory would be methodologically defined. Such an eventuality occurred in the middle third of the twentieth century with the rise of molecular biology.

It was in reaction to the rise of molecular biology and its effects on the whole of biology that Barry Commoner a half-century ago offered to *Science* a lecture he had previously given in which he bemoaned the widening bifurcation he was witnessing between what he called "traditional biology," including taxonomy, morphology, physiology, and evolution, and "modern biology," which is founded on physics and chemistry.[30] Commoner called attention to a house divided against itself in hopes of repairing family relations under one roof. He called for a collaborative relationship that might enrich both sides while saving traditional biology from the oblivion to which it was being consigned by increasing numbers of biologists. More and more biologists were of the opinion that science had collapsed any and all meaningful distinction between living things and non–living things and were treating biology as nothing more than an "unresolved form of chemistry and physics." Commoner wrote, "[A]s soon as an interesting and important biological problem becomes susceptible to chemical or physical attack, a process of alienation begins, and the question becomes, in the end, lost to biology.... [N]ow biochemistry and biophysics have reached deep into the core of biology — and the question arises as to how biology will sustain this more penetrating attack."[31] At stake for Commoner was an understanding of life that appreciates the uniqueness and complexity possessed by living forms above and beyond any of those biological elements or organic processes yielding to a physio-chemical determination.

The rift Commoner hoped would heal has, if anything, only deepened in time. Today, for example, high school students taking the Scholastic Achievement Tests have got to choose on the biology test between an emphasis on molecular biology or an emphasis on ecological biology. This distinction is not exactly Commoner's, but neither does it fall beyond the thrust of his reflections. More telling from our point of view, perhaps, is the distinction drawn by Lewontin between "manipulators" and "observers." Lewontin writes, "The division between those who try to learn about the world by manipulating it and those who can only observe it has led, in natural science, to a struggle for legitimacy.... In biology the two camps are now generally segregated in academic departments.... But," he points out most perspicaciously, "the battle is unequal because the observers' consciousness of what it is to do 'real' science has been formed in a world dominated by the manipulators of nature."[32] As usual, Lewontin's insight cuts to the core. Commoner's defense of "traditional biology" attempted a holding action against forces made superior by their alliance with the larger, dominant scientific community — a community whose professional culture has been shaped by modern (astro)physics.

The scientific culture of the manipulators understands science proper in terms already fully familiar to us. According to this understanding, science at its proven best realizes itself in certain, predictive knowledge: science consists in causal knowledge that it obtains by manipulatively interrogating nature's objective reality in order to effect "if ... then" predictive formulae of an instrumental character. This is the exemplary form of science; it is the scientific model arising with Newtonian science and sustained by relativity physics. The model itself and the scientific culture attached to it are rightly designated "positivistic," a term loosely derived from Auguste Comte's nineteenth-century sociology, which promoted "social physics," human progress, and enlightenment science, all within the framework of *Positive Philosophy*.[33] Such understanding later found programmatic formulation in the general dictates of the

Vienna Circle of Logical Positivism, which in the 1920s undertook to unite the epistemological grounds of all knowledge upon the model of classical physics.[34] We alluded to this group in the last chapter, and here we want merely to note that in biology, or in the study of organic phenomena in general, following the positivistic model of classical physics amounts to what is called *reductionism*. The manipulators in biology are well acculturated along the lines of positivism and are reductionistic in both their practice and their demand that the study of all things biological analytically reduce to physical explanation and lawful prediction.

Accelerating that acculturation in the middle third of the twentieth century were men who had been trained as physicists and who left physics to help found molecular biology. Bringing to the study of organic change a decidedly manipulative bias, Erwin Schrödinger and Max Delbruck were among the more renowned of those figures who forsook microphysics to plow new scientific fields. Schrödinger's short work *What Is Life?* was widely influential among that early generation of molecular biologists. It was not long before genetics took the throne in the kingdom of molecular biology, the coronation coming, of course, with Watson and Crick's earthshaking discovery of the structure of deoxyribonucleic acid, or DNA, in 1953. This discovery launched molecular biology as a whole along the spectacular career trajectory it enjoys today. If any substantiation of the reductionists' program were needed within biology, the discovery and implications of the double helix certainly were it. So, too, if the larger scientific culture of the manipulators needed any boost in the arm, Watson and Crick provided it. Signaling the heyday of the manipulators' imperious rise within the biological community, the discovery of the structure of DNA hastened the impending divorce Commoner later bewailed in the early 1960s.

So portentous was the discovery of the double helix that its impact on science and society cannot yet be assessed. Our concern with mastery of nature fortunately limits our interest in that impact, making it manageable. At this point, however, we postpone any consideration of those genetic manipulators who seek mastery of nature. In the next chapter we will critically discuss their identification of instrumental adaptation with maker's knowledge as well as their designs on human evolution. Before going there, however, we need to look at the perspective of those scientists belonging to what Commoner called "traditional biology." We want to address living things from their standpoint, focusing on what makes them "observers," to keep with Lewontin's distinction. Those whom Lewontin has in mind when he speaks of observers may come from a wide variety of fields and subfields, but generally they approach the study of organic phenomena with an understanding radically different from that of the manipulators. Without denying the significance of what the manipulators can do, these biologists have a larger, more inclusive vision of what a scientific biology needs entail. They belong generally to the tradition to which

Darwin himself adhered as a field observer who took his orientation from other scientific observers; from paleontologists and geologists, whose work rested on empirical grounds quite different from that of maker's knowledge.[35]

Modern scientific biology crystallized fully in the middle of the twentieth century in the confluence of Darwin's evolutionary theory, based on extensive fieldwork, genetic experimentation, sprung from Mendel's early work on biological variation, and population dynamics. The integration of these three areas of theory and research in the "synthetic theory" is what biologists loosely call neo-Darwinian theory, or simply neo-Darwinism.[36] Our discussion in the next section primarily will follow the compass of neo-Darwinism, as we explore those portions of the synthetic theory's basic conceptual territory that are directly relevant to our critique of mastery and that will guide discussion to follow in later chapters. Our aim essentially will be to identify in novelty and contingency constituent elements of organic change and organic development that elude a manipulative net. Our point ultimately is that biologists may manipulate where they can, but where they cannot, they perforce become observers. Anticipating this argument, we turn to a thought Gillispie offers us in his *Edge of Objectivity*, where he addresses Darwin's opening the door to a scientific biology:

> In mechanics, Galileo achieved objectivity by accepting motion as natural, and considering its quantity as something to be measured independently of the moving body. This he accomplished by treating time as a dimension, after which translational motion is no longer taken as metaphorical change. In Darwin ... natural selection treats in the same way that sort of change which expresses itself in organic variation (*the motion of species, so to say, through historic time*). Instead of explaining variation (or motion), he begins with it as a fundamental fact of nature.... This is what opened the breach through which biology could follow physics into objectivity [italics added].[37]

We want to question the validity of this parallel between physical motion and biological variation — or, more particularly, the parallel between the time of physical motion and the time of organic change.

Storytelling in the Natural Life-World

Gillispie correctly points out that Darwin simply took biological variation as a fact. Not until Mendel's work in genetics would evolutionary thinking obtain a proper explanation of variation. Yet variation itself was the critical axis around which Darwin turned the idea of natural selection, as this idea gave a "mechanistic" explanation of the material changes wrought by evolution. In the judgment of more than a few biologists, however, Darwin's real revolution lay not in the idea of natural selection but in his focusing evolutionary thought on the supreme significance of individual variation within a species, or, more precisely, within a local population of some species of plant or animal.

Evolutionary thought had enjoyed considerable currency in nineteenth-century Europe before Darwin's publication of *On the Origin of Species* in 1859,[38] and Darwin's argument that all living species descend from a common ancestor (descent with modification) by and large gained fairly wide acceptance, vociferous religious opposition notwithstanding. Evolutionary thinking before Darwin, however, had been bound to essentialist understanding focusing on the organic "types" or "forms" long studied in morphological change and schematized by paleontologists. These forms seemed hypnotically to exercise a Platonic effect on much if not most evolutionary thinking before Darwin's time.[39] Darwin, while not entirely consistent in his own thinking, radically transformed evolutionary thought by transferring the question of evolution from the domain of types to the domain of populations, as populations are understood to be groupings consisting of individuals possessing unique characteristics. Members of a species resemble each other and interbreed, which gave Darwin a clear sense of the larger species unit to which populations themselves belong, but it is on the individual variations within species' populations that Darwinian natural selection works. In other words, individual variations provide the specific material upon which nature acts somewhat like a non-selective breeder—blind to purpose, yet lawful in the cause of variational selection. As Darwin put it:

> Let it be borne in mind how infinitely complex and close fitting are the mutual relations of all organic beings to each other and to their physical conditions of life. Can it, then, be thought improbable, seeing that variations useful to man have undoubtedly occurred, that other variations useful in some way to each being in the great and complex battle of life, should sometimes occur in the course of thousands of generations? If such do occur, can we doubt (remembering that many more individuals are born than can possibly survive) that individuals bearing any advantage, however slight, over others, would have the best chance of surviving and procreating their kind? On the other hand, we may feel sure that any variation in the least injurious would be rigidly destroyed. This preservation of favourable variations and the rejection of injurious variations I call Natural Selection.[40]

Variational evolution has arisen as an expression more telling of what natural selection is all about than "natural selection" itself.[41] Modern genetics, beginning with Mendel, emerged from an interest in understanding individual variation upon testable terms of experimentation, and what would crystallize towards the middle of the twentieth century in the "synthetic theory" was the convergence of Darwinian understanding, genetics, and population dynamics in the study of organic change. Variational evolution was the meeting place in which these fields of study joined.

Approaching evolution from below, as it were, genetics addresses the source of what makes each living thing biologically unique among its population or species. Sexual reproduction is the primary generator of variation for sexual species. The chemical composition of human genes differentiates our species from all others, of course, but it also uniquely particularizes each person according to the conception of DNA arising when sperm and egg embrace. To all intents and purposes, the sequencing of genetic material along a DNA molecule allows variation to be virtually infinite, which is why no two members of any species are ever identical.[42] Sexual reproduction does not straightforwardly explain everything on this level, however, for variation also can spring from the random alteration of genetic material, which we commonly call mutation. Various kinds of mutation exist, the most surprising and exotic belonging to the transfer of genetic material across diverse species through some kind of interactional transfer. Cats and baboons, for example, having had no common ancestor in seventy million years, now share an identical piece of DNA that could have been given to both by a mosquito.[43] Mutation more commonly occurs within a species population, and most ordinarily with the replication of a DNA molecule during the life cycle of the cell. Such changes originate novelty in what amount to "typing errors" in the transmission of genetic information. Such novelty, in turn, is entirely random in manifesting arbitrariness in the relationship between a particular organism and its environment. Most mutations for this reason quickly become failed experiments in living, though survivors can gain some added environmentally specific advantage.

The genetic uniqueness of each living thing is the product of its genes and is synonymous with its "genotype." Genotypes are at the ground level of evolution inasmuch as they manifest the genetic fact of variation. Yet genotypes themselves do not evolve, and at death an individual's genotype is essentially identical with what it was at birth.[44] Between birth and death there is a great deal of non–genetic change that each individual member of a population goes through, but this change belongs to the phenotype, not the genotype. The phenotype is the individual organism whose individuality is a product of genotypic material and interactions with the environment, those interactions being ongoing, changing, and to no small extent situationally contingent. The phenotype embodies what is unique about each individual organism relative to all other members of the population to which it belongs. Phenotypic change of an individual entails genotypic material without reducing to those genetic elements. The crucial factor defining why or how the phenotype transcends such reduction is its life-long interaction with its environment.[45] Or, as Lewontin articulates the nature of the phenotype most sharply, "The internal and the external, what we now think of as the gene and the environment, meet in the organism."[46]

Darwin's understanding of variational selection identifies evolution in terms of phenotypic success or failure.[47] Success belongs to those phenotypes whose variational characteristics enable them to deal most effectively with the demands and/or challenges of the environment. Those variations able to deal most successfully are able to survive and reproduce, and within and through heritable material, reproduction passes along those characteristics that favor survival. Those traits are the stuff of adaptation. As Simpson puts the point, "[A]n adaptation is a characteristic of an organism advantageous to it or to the conspecific group in which it lives, while adaptation or the process of adaptation is the acquisition within a population of such individual adaptation."[48] Adaptation and evolution thus respect populations of individuals as these individuals compete to survive.

While the phenotype is the meeting place of interactions between the genotype and its environment, the population itself is the meeting place among phenotypes both in their competition with each other and in their reproductive success or failure. This point returns us to Darwin's insight into evolution as something other than a changing of forms. Mayr summarizes the matter as follows: "Species and populations are not types, they are not essentialistically defined classes, but rather are biopopulations composed of genetically unique individuals.... It was Darwin who introduced this new way of thinking into science. His basic insight was that the living world consists not of invariable essences (Platonian classes), but of highly variable populations," and, as he continues, "It is the change of populations of organisms that is designated as evolution. Thus, evolution is the turnover of the individuals of every population from generation to generation."[49]

Here we have in outline the rudiments of the synthetic theory, and the question to be asked at this point is this: what is there in the causality of adaptation and evolution that places evolutionary scientists in the role of observers? Critical to that causality is the variation of the genotype, considered random insofar as sexual reproduction and mutation introduce new variations. Mutations themselves may be failed experiments in living in the vast majority of cases, but they do enhance the introduction of chance and novelty into evolution on the ground floor. Natural selection then goes to work lawfully on the phenotypes, whose survival couples basic genotypic variation with the random elements and vagaries of circumstantial interactions between the genotypes and their often changing environments. Novelty, uniqueness, and contingency thus mix with genetic material in the unfolding success or failure, life or death, of the organism. The critical point for us here is that the causality implicit in such organic change and development is temporal; it abides in time. The laws of evolution may themselves be considered timeless in their invariant application, their causality being always the same regardless of time or place. The material of variation upon which and through which those laws work, however, is saturated with time, or more particularly, with life-time.

Hyphenating life-time here bears no affectation as it intends merely to distinguish organic time from the time discussed in connection with motion and dynamics, which is the time of physical motion captured in magnitudes of velocity/momentum/acceleration. We need, then, to be clear about what this life-time refers to, or entails, in connection with organic change. Three reference points obtain in this idea. First, life-time refers to the lifetime internal to living things. Such lifetime manifests itself in the developmental stages organisms "go through" in the passage from birth to death. This lifetime is the stuff of maturation and aging. Second, the idea of life-time regards the dynamic interrelationship of the genotype and its environment. This lifetime abides in the interactions affecting the nature and changing character of the phenotype as the phenotype unfolds, or less developmentally phrased, as environmental interactions condition the organism sequentially. This lifetime is the stuff of biography. And last, but certainly not least, life-time refers to the passing of generations, each generation disappearing into the next in what sometimes is called "intergenerational time." Most generally, of course, this lifetime is the stuff of species adaptation and evolution itself.

Darwin certainly knew full well the significance of time to the material of his observations. Speaking of evolution as ongoing in the grand sweep of geological ages, he wrote that it "may be said that natural selection is daily and hourly scrutinizing, throughout the world, every variation, even the slightest; rejecting that which is bad, preserving and adding up all that is good; silently and insensibly working"; he then remarks, "We see nothing of these slow changes in progress, until the hand of time has worked the long lapse of ages...."[50] Or, as Ernst Mayr one and a half centuries later directly connects the observational standpoint of evolutionary biology to the impact of Darwin's work, "As far as science is concerned, the acceptance of evolution meant that the world could no longer be considered merely as the seat of activity of physical laws but had to incorporate history and, more importantly, the observed changes in the living world in the course of time. Gradually, the term 'evolution' came to represent those changes."[51] Inasmuch as the causality of adaptation and evolution abides in life-time, the standpoint of the evolutionary biologist is historical precisely because he or she is trying to account for what caused some particular phenomenon under study to have come about or emerged through time. Considered as *observers*, then, evolutionary biologists are historians of a scientific sort.

It is not simply the grand sweep of evolution that brings history into view in evolutionary biology. The primary interest of the evolutionary biologist lies in explaining this particular adaptation or that, how this or that particular structure or function has come about, all of which entails history.[52] Such a historical approach to biological study, not surprisingly, brings with it a form of scientific explanation different from what has been seen and discussed in connection with the maker's knowledge of the scientific tradition launched by Newtonian science. More directly, determining the causality of evolutionary phenomena methodologically reverses the considerations of cause and effect found in physical science. Simpson puts the point this way: "The most frequent operations in historical science are not based on the observation of causal sequences—events—but on the observation of results. From those results an attempt is made to infer previous causes," and, then further: "Prediction is inferring results from causes. Historical science is largely involved with quite the opposite: inferring causes (of course including causal configurations) from results."[53]

Methodologically speaking, in such biological understanding what is *given* is the effect, not the cause. Explanation is post–dictive, but unlike the post–dictive nature of quantum "understanding," the character of biological explanation rests on no experimental follow-up. The credibility of the explanations we find in evolutionary biology may find support in experimental work, but far from reducing to experimental proof, such credibility derives from a very different kind of source. Explanation is bound to the historical record and all related empirical data, and the persuasiveness or convincing power of evolutionary explanation lies exactly in the plausibility and coherence of the causality reconstructed through historical investigation. Scientific explanation of this sort comes in the form of a story that, in true historical fashion, plots its meaning, or, in this case, its causality, in narrative form. Goudge, a philosopher writing on evolution, is clear and concise on this matter:

> [Understanding] may involve reconstructing certain of the events, selecting those which vitally affected or foreshadowed later developments, arranging the events in a significant order, etc., so as to tell a "*likely story*." The aim is to make the sequence of events intelligible as a relatively independent whole.... [The] explanatory pattern we are considering forms a coherent or connected narrative which represents a number of possible events in an intelligible sequence. Hence the pattern is appropriately called a "narrative explanation" [italics added].[54]

Narrative explanation traces causality backward from the given effect by way of records, fossil and other, and then, turning matters around again, interpretively reconstructs the temporal sequence and interaction of causal conditions leading up to, or effecting, the given. Narrative explanation in this way provides a story-form of causality. Hence, when we speak metaphorically of the "story of evolution," we actually speak redundantly inasmuch as evolution is, in the most meaningful sense, already a story.

Evolutionary biologists' explanations are necessarily narrative because the causality nature employs in evolution requires the form of a story. That story is unfinished; evolution is continuing, just as organic change of all sorts occurs everywhere and at all levels of nature without surcease. The irony, then, is that while evolution goes ahead, our "grasp" of it generally can only follow upon this or that particular facet of what evolution already has wrought and left behind, as it were. As science philosopher Michael Scriven wrote some time ago in a widely influential article, "[Evolutionary] theory shows us what scientific explanations need not do. In particular it shows us that one cannot regard explanations as unsatisfactory when they do not contain laws, or when they are not such as to enable the event in question to have been predicted."[55] If the symmetry of explanation and prediction ran into trouble in quantum mechanics, here, given the thrust of Scriven's point, it simply goes by the board. Quantum theory may fall short of knowing the future experimentally, yet it can at least mathematically formulate probabilities that are bound and determined by the experimental situations themselves. Evolutionary theory, on the other hand, generally writes off prediction and says little or nothing about the future.

Under these circumstances biological thinking sometimes elasticizes the meaning of "prediction" to entail "retrodiction," which amounts to predicting what we will find out about the undiscovered past if our theory is correct. Stephen Toulmin some time ago played with this idea in his *Foresight and Understanding*,[56] and quite recently Niles Eldredge endorsed retrodiction while arguing persuasively that Darwin himself employed the idea strategically throughout his groundbreaking book on evolution.[57] Identifying prediction with retrodiction nevertheless only outlines in the manner of negative space what evolutionary understanding cannot do: it does not enable us to predict, to say knowingly before the event, what will occur in the future. Although many kinds of biological phenomena may be predictable, Goudge points out that "[t]he factors [an evolutionist] is able to specify are insufficient to allow a standard 'if ... then ...' proposition."[58] And as Lewontin speaks more specifically of similar limitations running across various levels of our "grasp" of organic change, "Even if I knew the complete molecular specification of every gene in an organism, I could not predict what that organism would be.... Moreover, curiously enough, even if I knew the genes of a developing organism and the complete sequence of its environments, I could not specify the organism."[59]

For scientific biology to admit such non–predictive, historically observational understanding is anathema to those for whom the *sine qua non* of science is that form of understanding first realized in Newtonian science. From such a positivistic perspective, evolutionists ultimately are *only* historians, and, as mere observers, they do not belong in the company of bona fide scientists who, like molecular biologists, aim for "if ... then" manipulation in their work. This disparagement or dismissal of the evolutionists' form of biological explanation bifurcates biology, establishing the departmental encampments of which Lewontin spoke and which Commoner lamentably addressed in his *Science* article. In science laboratories experimental manipulation can grasp many elements of organic change. But in this study we will agree with the observers that nature also has a laboratory, its own laboratory, where it works out the causality of organic change on its own terms. In that laboratory genetic factors, population factors, and environmental factors conspire circumstantially and interactionally in immensely complex and singular ways to make significant organic change novel, unique, and contingent.

Different biological "observers" articulate this large point differently. Simpson spells out the *modus operandi* of nature's laboratory this way:

> The regulation, which seems surely to be a determination, is by forces that change continuously in intensity, direction, and combination, and that produce quite different results in different instances. These forces are themselves interwoven with the historical process and subject to historical causation.... Scientific prediction depends on recurrence and repeatability. Precise prediction of unique events is impossible either in practice or in principle. Historical events are always unique in some degree, and they are therefore never precisely predictable.[60]

In a similar vein Stephen Jay Gould writes in his inimitable manner of infusing scientific judgment with personal sensibility as follows: "So, ultimately, the question of questions boils down to the placement of the boundary between predictability under invariant law and the multifarious possibilities of historical contingency.... I envision a boundary existing so high that every interesting event in life's history falls into the realm of contingency."[61] Goudge approaches such matters more austerely by addressing the non–linear nature of the relevant causality. He writes, "For evolution is a historical process, and hence some of the conditions which enter into its causal determination are cumulative. Each phase of the process adds fresh conditions to those already at work, so that a return to the precise set of conditions which prevailed at the outset of the process is impossible."[62] Whereas in classical physics the movement of understanding from present to future (cause to effect) and present to past (effect to cause) entailed no significant difference in the timelessness of causal understanding, in the causal dimensions of organic change life-time is at once real and meaningful. Life-time becomes telling.

At this juncture, Gillispie's earlier parallel between the time of physical motion and the "motion of species through historical time" appears most ironic. Galileo was not a storyteller; nor was Newton. Modern astrophysics fixes time in temporal magnitudes and realized causal laws of motion, making explanation and prediction logically correspondent in strictly deterministic "if ... then" form. The physical knowledge realized in those causal laws certainly did not eliminate time in the actualization of physical events, but it did render time nonessential, irrelevant, indeed practically nonexistent in the grasp of understanding. Consider now in the place of Laplace's classical Newtonian seer, a superhuman evolutionist—a superhuman seer of all living things for whom the causality of all organic change would be knowable. In contrast to the Laplacean seer's ability to see all things past and future in the knowable present, this biological seer's superhuman status would spring not from timeless knowledge but rather from his or her having virtually everlasting life here on earth. This seer would have to possess some preternatural ability to live beyond the ending of all other life. He or she could patiently unearth and study all past life as this accumulated, and he or she could patiently interweave the countless, unfolding stories arising with such continuing study. Most critically, however, this seer would have to live until the end of time, or at least to the ending of all life itself, in order to know all of what might yet appear along unpredictable pathways of unique and interactionally contingent "events" to come. For him or her to know the story of life, all organic change would have to have reached its conclusion and be entirely complete. Our super-human biological seer thus would manifest in his or her own longevity exactly what understanding the "motion of species through historical time" entails: namely, life-time.

The time of organic causality and the time of physical causality seem to enjoy comparable status in Gillispie's rendering, and such comparability is precisely the problem there. Whereas in classical physics one could lay one's hands on time and the future in the predictive measure of motion, life-time is different. Lifetime is not itself causal. Rather, lifetime is the dimension in which and through which the biological causality of organic change works via uniqueness, novelty, and interactional contingency. Lifetime may amount to nothing meaningful in those cases where an aspect or instance of organic change is controllable. However, in those cases where the causal agents of novelty and interactional contingency elude the manipulative grasp of experimental science, life-time comes to govern the realm of appearance. In those circumstances, scientists are not manipulators; they are observers. They are not masters; they are storytellers.

Storytelling will become increasingly thematic in the remaining chapters of this study as we turn to various case studies of our instrumental interactions with biological and ecological phenomena. We will argue that within the natural life-world in which we are enmeshed and to which we are inextricably bound, our inability to control organic change makes us storytellers in the same way it makes our knowledge uncertain. In point of fact, as a mode of scientific understanding, storytelling properly aligns our inability to control natural processes with the uncertainty we experience in our instrumental interaction with nature. As we unfold this argument in the following chapters, we will contend with the two major programs proposing mastery over these very same natural processes. Not content with the ungoverned and piecemeal project of mastery materialized in the industrial revolution, and in some cases being critical of that revolution's haphazard disregard and blind abuse of nature, these programs are just that—programs, designs, for human dominion. They are programs that see fit expressly to identify instrumental adaptation with complete instrumental control of one kind or another. Quite simply, in these designs on dominion, instrumental adaptation comes to mean our exercising self-conscious mastery over virtually all natural processes.

Contemporary projects of mastery come in two prominent designs. One design is genetic and springs from molecular biology; the other design is planet management, which draws upon systems science. Instrumental adaptation in both cases means designing the future and controlling our evolution in some sense. Genetic masters and planet managers generally share the belief that, with the appearance of humankind, evolution unwittingly or not gave rise to a species that has developed to the point of transcending most of the limits placed on all other life forms. Humankind possesses awesome control capabilities as well as the self-consciousness necessary to put these capabilities to work in the design of reality itself. Each perspective on design and control bears meaning particular to the field of scientific study from which it springs. When those promoting genetic engineering have the floor, they identify the design of human evolution with maker's knowledge. Indeed, they will make nature over, pure and simple. When those promoting planet management have the floor, they identify the design of our interactions with our biological and physical environments with systems simulation on supercomputers and feedback control. They will guide the future of this planet Earth. Neither perspective denies the other's legitimacy, and both can readily cohabit the consciousness of visionary masters like Michael Kaku, the cofounder of string field theory and a widely known promoter of science, who speaks of the biomolecular revolution and the computer revolution as two of the three pillars of our coming conquest of everything in sight, ultimately including the sun and beyond.[63]

Some formal design for human dominion seems an almost inevitable offspring of any programmatic union of the idea of instrumental adaptation with the illusion of mastery. Molecular biology and systems science do not themselves produce such a marriage, but they offer design tools that can, and do, outfit would-be masters. Those tools do not supplant the conventional instruments of the modern Prometheus; they simply go beyond them in scope and ambition in one way or another. What remains the same in any case is the illusion of mastery. Our argument in the following chapters will aim to puncture that illusion while unfolding the role of storytelling in the land of sometimes and perhaps.

5

Elusive Elements of Organic Change

One would be hard put to find a scientific discovery to rival that of James Watson and Francis Crick's determining the structure of deoxyribonucleic acid (DNA) in 1953. Their achievement possessed an immediate significance at once culturally symbolic and practically prodigious in its promise. Telling in both respects is the proclamation Watson recalls Crick making at the time to those sitting in their favorite pub: he told "everyone within hearing distance that we had found the secret of life."[1] If Faust and Prometheus could ever have a meeting place outside of myth, it lies there in that public announcement. And in this particular case it is actually quite difficult to extricate science from myth, for the idea of life's having a secret itself partakes of the very stuff of myth. Life's secret is the key to everlasting life; it fulfills the wishful pursuit of immortality. The quest for immortality in Western thought goes all the way back to the Sumerian epic of *Gilgamesh*, only to recur in various cultural renderings throughout the following four and a half millennia. Faust and Prometheus can both be located in that genealogy, broadly conceived, and Crick's claim certainly shares some of the genetic material to be found there.[2]

Contemporary associations of immortality with genomic science operate on different levels of discussion. High drama and media attention flourish where questions concerning our "playing God with DNA" attract philosophical and/or religious opposition. The millennium's turn saw this kind of discussion take place at the University of Pennsylvania's museum, where a conference on immortality was held among microbiologists, ethicists, and theologians. The last two groups came from many quarters but found common voice in opposing the possibilities and implications of genomic science raised by the matter-of-fact scientific assertions they heard. According to science writer Nicholas Wade, "The scientists at the meeting said, in effect, that life span was clearly a phenomenon under genetic control, so they expected to be able to increase it just as soon as they laid their hands on the right genes."[3] *Life span* befits scientific sensibility far better than "immortality"; it sounds more reasonable and bears quantifiability. Experimentation along these lines on laboratory animals is expressed in terms of a life span that, in the case of mice and roundworms, can be altered within short periods of generational measure. That said, the search for immortality still echoes loudly in the cavernous promise of the scientific manipulation of our genes. It resounds not only directly, in discussions of possibly tripling or quadrupling our lifetime, but also indirectly in hopeful expectation of the possible elimination of disease. These matters belong, of course, to a larger discussion of the genetic manipulation of life.

Mastery of the organic world seems very much on the march today as biotechnology advances in so many directions. Genetically modified plants and organisms are proliferating in widely disparate applications: transgenic plants are providing microbial proteins to effect animal vaccines; engineered microbes are working in various bioremediation projects, oil spills being the most widely publicized; genetically altered fish are serving as biological sensors to monitor mutagenic agents and water contamination; and transgenic plant species are being continually introduced and harvested for resistance to pests and for their nutritional benefit, while others are being developed for pharmacological value.[4] Just cataloguing such examples could be a full-time job. Less developed is the potential of genetic design for human beings, where the stakes of manipulation entail moral and political problems enormously compounding those of a research and experimental nature. But the real potential of genetic design for human beings is huge and widely touted, the ringing alarm bells notwithstanding. Indeed, many new masters of nature see in genetics unprecedented powers relegating all previous advancements of modern science and technology altogether secondary in significance. What they speak about is nothing short of *enhancing* the constitution of nature itself.

We want to explore the thinking behind such proposals of mastery, and especially those targeting our own species, in the first part of this chapter. In the identification of instrumental adaptation with genetic engineering we will find that the illusion of mastery carries with it certain dangers. After critically discussing genetic design in general connection to our own species' future, we then will focus thematically on certain aspects of disease that suggest most strongly, if not certainly, that struggle will forever mark our relationship with bacteria and viruses that sicken and kill. Talk of human immortality brings nothing but laughter among the microbial populations inhabiting this planet. Those populations have the knowledge and wherewithal to make us storytellers even as we advance our scientific knowledge in manipulative fashion.

Instrumental Adaptation as Genetic Design

When Francis Bacon spoke four centuries ago of "enlarging the bounds of human empire to the effecting of all things possible," he had in mind the material provided us by nature. In Bacon's program nature itself cuts the causal keys to maker's knowledge, and once science comes into possession of these keys, humanity can unlock nature's secrets for human use and benefit. Today's masters in molecular biology generally follow Bacon's lead in their experimental reductionism and in their interest in pursuing knowledge for human well-being. Yet, in their hands, the effecting of all things possible means something essentially new in the project of mastery. Organic nature now becomes subject to the kind of manipulation that may manufacture life itself, and for the first time the very constitution of nature appears fully open to human judgment and alteration. The unprecedented issues arising with such power concern whether or not nature is in this or that respect suitable for human need or human use, and much more profoundly, whether or not we should simply proceed programmatically to remake nature, introducing improvements as we see fit.

The molecular biologist Lee Silver has entitled one of his works *Challenging Nature* and another *Remaking Eden*. Both titles capture the thrust of much thinking in genetic engineering. Or, as David Garreau quotes Don Kash, one of the world's leading experts on instrumental innovation: "'Our capabilities are now so great that nature is not something you study in order to figure out how to make things work.... Nature is something you study to figure out how to engineer and change it. Including human nature.'"[5] It seems nature no longer meets our needs or desires; or, if this formulation seems too strong, nature suffers many inadequacies that need remediation from humanity's point of view.

Throughout modern time the project of mastery always addressed nature from humanity's point of view. Until now and with few exceptions, however, nature had had about it some essential fixity in form and content. Nature possessed, if you will, a *nature*, and the nature of nature went unchallenged simply because nature and its workings were thusly constituted. Such understanding goes out the window for the new masters who address things from the standpoint of molecular design. From their perspective, we hamstring our power if we even think of nature as fixed in its stuff and ways. They are not altogether wrong in this perspective, for the capabilities lying within reach of genetic manipulation are prodigious beyond any preexisting measure. The point here, however, is that once the nature of nature is thought subject to human domination, our very designs on nature themselves tend to become unlimited.

Jeremy Rifkin has altered Joshua Lederberg's idea of *algeny* to suit this way of viewing nature as altogether subject to human invention and perfectibility, arguing that it is a way of perceiving nature as well as acting on nature that is itself novel.[6] From this revolutionary point of view, in principle at least, nature does not appear to forbid our rendering all things artificial, should we so choose. Against our will, nature has no voice.

Such scientific presumption often finds bold articulation. Lee Silver writes in his *Challenging Nature*, "Slowly, inevitably, human nature will remake all of Mother Nature in the image of the idealized world that exists within our own minds—which is what most people really want subconsciously."[7] Silver sees this remaking as continuous with what has gone before, but, important industrial synthetics notwithstanding, the expansion of human artifice constituting the making of the modern world assumed natural boundaries that genetic designers simply do not recognize.

Consider what Michael Fossel writes in his *Cells, Aging, and Human Disease*: "Recent work in cell biology and genetics has suggested that at the most fundamental level, not only are age-related diseases the outcome of cell senescence, but our growing understanding of cell senescence, senescence-associated gene expression, and telemerase biology will soon allow us to prevent, ameliorate, or to some degree even reverse such pathology," and as he concludes, after having considered the problems to such progress posed by human ignorance, poor thinking, and the low supply of human kindness, "[i]t may, at times, appear impossible that we can ever overcome such problems. It may appear equally impossible that we can ever overcome aging and the abundant diseases in which aging is manifest. Our limits [however] are not set within our world, nor in any heaven we may conceive. *Impossibilities are defined by people, not nature* [italics added]."[8]

Although not everyone promoting genetic engineering would subscribe to the categorical implication of Fossel's meaning, most would certainly agree that human beings can and do throw up limits where nature may not. Genetic engineers see cultural mores and existing laws sedimenting long-held Judeo-Christian values as the primary obstruction to our open-ended progress on the genetic front. Whereas Asian cultures currently seem generally more receptive to the myriad research directions and practical possibilities of genetic engineering,[9] Western peoples and their governments today tend still to resist the lion's share of the proposed human applications of genetic knowledge believed soon to be within our grasp.

Silver's *Challenging Nature*, which is subtitled *The Clash Between Biotechnology and Spirituality*, carries on a sustained argument against those opposing biotechnology, be they Christian opponents, whom he deems conservative, or naturalistic naysayers embracing "Mother Nature," whom he sees occupying the cultural "left."[10] In both cases Silver sees spirituality of some sort either denying the propriety of our pursuing such instrumental knowledge or limiting the application of genetic knowledge to narrowly circumscribed areas of permissible use.

According to a 2002 poll, while only 20 percent of Americans believed it acceptable to use genetic engineering to design the traits of the unborn, just under 60 percent thought it nevertheless permissible to employ biotechnology to eliminate any heritable diseases detectable in the DNA of an embryo.[11] This position corresponds loosely with that taken in 1997 by the Council of Europe, which tightened yet further the conditions of permissible genetic intervention by tethering to somatic intervention disease prevention and medical therapy.[12] As distinct from germline gene therapy, which targets gametic cells for heritable transmission from one generation to the next, somatic gene therapy is limited to the individual receiving it and confers nothing heritable upon the generation(s) to follow. The position assumed by the American public and the Council of Europe seems broadly in keeping with the thinking of a wide spectrum of bioethicists who would disallow germline engineering while binding somatic therapy to matters of individual health and medicine. Such thinking meets the interests of those promoting genetic design for human beings only partway. For those promoting genetic design, binding genetic intervention to matters of health and medicine is still a bind. As important as medical intervention is, institutions are placing limits on what we can, or might, know and do. It is not nature, but people, who are defining impossibilities.

One way for the manipulators to counter such limitation is to deploy the term *disease* as a rhetorical strategy to open the range of the permissible. Fossel, for example, speaks of disease in this way where he calls aging a "pathology." His choice of words is anything but random, as genetic research on aging has shaped such meaning in the thinking of many molecular biologists. Guarente and Kenyon, two widely respected genetic researchers in the field of aging, have written in *Nature* that "[i]t is now widely accepted that the ageing process, like most biological processes, is subject to regulation and can be studied using classical genetics. When single genes are changed, animals that should be old stay young.... On this basis we begin to think of ageing as a disease that can be cured or at least postponed."[13] Stretching the meaning of *pathology* too far, however, can affront common usage and common sense. Hence, many of those offering genetic engineering as a choice technology for individuals and families to use as they see fit take a different route around the limitations on biotechnology posed by the criterion of medical therapy. Not wanting to elasticize the meaning of disease too far, the new masters often speak of the impossibility of making any strict or meaningful distinction between the genetic treatment of disease and genetic intervention for human *enhancement.*

Enhancing a human being and healing a human being do not appear to be the same thing at first blush. Enhancement appears to lie outside or beyond curing or healing; it is in some sense additional. Nicholas Wade, who at different times has written for *Nature* and *Science* as well as for the *New York Times*, and who is quite sympathetic to the entire cause of genetic engineering, gives us just this sense of difference where he identifies the kinds of things enhancement thinking surveys.[14]

> Assuming ... a method is invented for reengineering the human genome, it would be possible to go beyond the replacement of disease-causing gene variants and also insert genes that enhance desired qualities, such as strength, beauty, intelligence,

> and any other human attribute that has a genetic basis. A couple's children would be their own, but parents could select from among the most favorable gene variants in the human gene pool to enhance their children's qualities in ways they deem desirable.[14]

Any meaningful distinction between gene therapy targeting disease and genetic intervention targeting enhancement collapses, however, in the thinking of most who would "remake all of Mother Nature in the image of the idealized world that exists within our own minds," to recall Lee Silver's point. Silver himself argues that any genetic engineering for health or therapeutic purposes would constitute enhancement since the intervention goes beyond the genetic material obtained "naturally" from the parents. "Thus," he tells us in *Remaking Eden*, "in every case, genetic engineering will be genetic enhancement.... It is for this reason that I will use the term 'genetic enhancement' interchangeably with 'genetic engineering' and 'gene therapy.'"[15]

This conceptual conflation of therapy and enhancement identifies any and all designed change as enhancing, regardless of aim or consequence, and allows each one of us to be the judge: therapy is whatever one chooses to do to effect whatever one considers genetic improvement. Therapy or enhancement is really, then, a subject of individual choice. From this standpoint the appealing thrust of much enhancement thinking is also what separates the position of many contemporary genetic designers from the eugenics movement of the early twentieth century. Enhancement lies in the eyes of the beholder, and according to many of today's genetic designers, individual and family choices, not any governmental program, would identify genetic selection and democracy in one and the same universe of choice. Gregory Stock, as director of UCLA's Program on Medicine, Technology, and Society at the School of Public Health, put the point this way in his *Redesigning Humans: Choosing Our Genes, Changing Our Future*: "To see the future of GCT [genetic choice technology], we need to be more pragmatic and acknowledge that people want to be healthier, smarter, stronger, faster, more attractive. Enhancements are those modifications that people view as largely beneficial and that serve their goals. Virtually by definition, people seek such modifications."[16]

If human beings seek enhancement naturally, then the current cultural obstacles to the biotechnological design of human beings will surely prove temporary. Two generations ago the vast majority of North Americans doubtlessly would have found religiously repugnant and downright ghastly the idea of mixing sperm and ova in petri dishes and then implanting the fertilized ova in women's uteri. In the Rockwellian America of Beaver Cleaver's time, visions of *Brave New World* and thoughts of Dr. Mengele probably would have mingled horrifically in the imagined prospect of such fertilization. Today, of course, couples having difficulty with conception turn every day to this relatively safe and straightforward procedure of in vitro fertilization (IVF). And it is with IVF and this change of heart in mind that many see the strong likelihood, if not the absolute certainty, of genetic engineering's complete, unfettered arrival in the near future, as emergent technologies place within the scope of parental choice widening genetic options for their children. The hardback edition of Stock's *Redesigning Humans* originally was subtitled *Our Inevitable Genetic Future*. Or, as Ramez Naam articulates the point in species-specific terms, "the drive to alter and improve on ourselves is a fundamental part of who we humans are.... Many past enhancements that we now take for granted—from blood transfusions to vaccinations to birth control—were called unnatural or immoral when they were first introduced."[17] Presumably, our drive to enhance ourselves is one aspect of the nature of our nature that we will choose to keep should we happen upon an "enhancement gene."

If genetic engineering of human beings is in the wings just awaiting the prompts of continuing research and developing technologies, as most likely it is, IVF is the stage on which any anticipated dramatic entry of genetic engineering is likely to occur. In Silver's pointed words, "In a very literal sense, IVF allows us to hold the future of our species in our hands."[18] IVF provides direct genetic access to embryos, and since the first successful IVF human procedure in 1978, the uses of IVF have increased alongside advancing knowledge. Most importantly, IVF has allowed for preimplantation genetic diagnosis: first as a means of choosing the healthiest of embryos for implantation; then as a way of spotting genes bearing heritable diseases; and now as a means of searching out those genes that may increase the proclivities for developing this, that, or the other disease.[19] Continuing along this path could well mean arriving ultimately at germline intervention if engineering proponents have their way and presumably safe technologies have been found and approved for embryonic selection. Selection at that point will far transcend determining which of the embryos is best for implantation. It will mean choosing by biotechnological design the genetic makeup of the life to follow.

Existing design tools go back to the revolutionary work done with recombinant-DNA in the early 1970s, revolutionary work that lay the groundwork for cloning and led to the manufacture of the genetically engineered plants and animals that populate our planet so widely today. Restriction enzymes for cutting and splicing DNA sequences have been critical to the development of genetic engineering, and with this capacity to lay hold of genes, much attention has been given to issues of delivery, as the delivery of genes to target sites, typically by means of a viral vector, has involved problems of imprecision and safety. Vector problems have received attention across the board, especially in the case of human interventions, which have been limited thus far to somatic therapy and which have proved fatal in at least two cases.[20] Alternative methods of delivering genes to specific sites are being explored. One such possibility is "homologous recombination," a method employing a patient's own cellular operations to target the precise DNA site needing repair.[21] This method suffers its own shortcomings, however, and some genetic designers, including Stock, have favored using an artificial chromosome that might be inserted into the embryo without the problems and risks of vector delivery. According to this idea designers would be able ideally to provide an entire package of genes at once, and, if all the wishful hopes of the designers bear fruit, this technology would further enable them to engineer a genetic program for receiving updates as knowledge progresses, for pulling the plug on the genes if problems emerge, and for limiting germline therapy to a generation-by-generation decision either to continue with the same chromosome or to move on to a new and improved package.[22]

Whether or not the artificial chromosome could ever be all that its proponents want it to be, it points to crucial issues of choice and safety. Such technology addresses the possibility of closing down a germline intervention either within the lifetime of the enhanced individual or between that individual's life and the life or lives of those to follow in his or her genetic footsteps. Those mining the human genome have become increasingly aware of the complexity of the material they are handling. Much of what was recently thought to be "junk DNA," or DNA without any function, is proving to be something other than junk, and the interaction of genes in the genetic dimension of causality is very far from being readily determinable. Concerning genetic causality in general, Nelson Wivel, a former member of the Recombinant DNA Advisory Committee of the National Institutes of Health (NIH), wrote the following judgment of germline therapy[23]:

> The risks of the technique will never be eliminated, and mistakes would be irreversible. Germ-line gene modification will always be associated with the risk of unpredictable genetic side effects, and for this reason it never should be approved for use in humans. Whatever the mechanisms of review and approval, they are not likely to be fail-safe because it is not possible to guarantee safety and reproducibility in biological systems. Further, there is the ever-present potential for the delayed appearance of unpredicted side effects that could be passed on to [viz] future generations; for example, subtle adverse effects on the brain could appear many years after genetic intervention, and such effects might not be detected in animal models that were used to develop the preclinical data.[23]

Apart from any other questions of safety, the problems spotlighted by Wivel are the ones Stock and others hope to minimize or even sidestep via technologies such as the artificial chromosome.

Wivel's concerns respect the ultimate insufficiency of animal testing to insure human safety and, more profoundly, the non–linearity and/or temporality of much biological causality that escapes direct experimental determination. A given genetic effect may be polygenic, entailing a considerable number of interacting genes; or a given gene may be pleiotropic, having multiple effects. In such cases the relationship between cause and effect may not be experimentally determinable in any respect. Genetic causality may be altogether temporal in its nature, and the effects of the genetic intervention may occur only at different stages of the individual's development, and then not all at once. Changing environmental contexts can compound further whatever uncertainty abides in the knowledge initially accompanying an intervention. As Bioethicist Ronald Green warns, "Among other things, environmental influences can cause the suppression of some gene functions and the activation of others. But our new knowledge of genomic complexity tells us that genes and parts of genes interact with other genes, as do their protein products, and the whole system is constantly being affected by internal and external environmental factors."[24] Respecting both an individual's development and the intergenerational issue of heritability, this mixture of causal elements operates in the life-time of organic change, making prediction of biological effects problematic at the very least. Such causality renders the realm of appearance beyond experimental control.

Genetic designers do not deny the uncertainty inhering in the intervention they programmatically promote. All of them speak of there being some element(s) of risk. Estimating the probability of the unforeseeable (Donald Rumsfeld's unknown unknown), however, is an impossibility outside any calculation of risk, and the unforeseeable is itself the great biological danger where polygenic, pleiotropic, and environmental factors may organically conspire to turn wishful dreams into living nightmares. This critical point notwithstanding, the designers' willingness to accept risk is real, and it operates on two levels. Most designers seem to see risk-taking simply as our doing what we must do unavoidably. Ramez Naam informs us that it is only human to aspire to what lies beyond our immediate reach. Humankind never would have reached its present level of civilization without having taken risks; we owe everything to our curiosity, our bravery, and our "willingness to experiment with the unknown."[25] This willingness is precisely what Craig Venter, the entrepreneurial geneticist of Genome Project fame, sees in the activity of science itself. Responding to an interviewer's question about danger and caution in genetic engineering, Venter declares that science inevitably entails risks, and then he "honks and waves as he pulls out, a grinning Faustus racing out into the storm."[26] But Nicholas Wade would remind us, as he echoes Naam's naturalistic idea, that Faust is not a bad point of reference for thinking about the future of genomic science and the options lying before us:

> Like all great gifts, this one will come at a price. The price is that, in beginning to alter the human germline in the name of health, we will inevitably assume a broader control over it. Evolution having done its part, our childhood will have ended and we will indeed hear the voice of Mephistopheles. But Faust heard a good angel too; he just made the wrong choice. Having come so far and gained our first glimpse of the human genome sequence, will we really declare it too dangerous to handle? If that were our nature, we would never have risked leaving the familiar African savanna in which we arose. We can never return to its confines. We can never turn back.[27]

It seems that, from the genetic designers' point of view, science essentially manifests the nature of our nature, which is to take risks in order to advance and develop. As would be the case with an "enhancement gene," if scientists should come upon a "risk-taking gene," so, there too, they will need to place markers signifying "thou shalt not trespass here." We need, after all, to protect the genetic root of instrumental adaptation.

The second level on which designers address risk-taking speaks to the practical problem of handling what is at least potentially unknown. As noted earlier, inasmuch as animal testing and computer simulations are necessary but insufficient experimental approaches to determining human safety, designers seek to circumvent the conclusion to a possibly bad interventionist narrative by practical means. Ronald Green is one of the leading bioethicists embracing the promise of genetic engineering, and he is also one of the most cautious. One of his four guidelines for navigating the uncharted waters lying in front of us specifies two critical conditions for germline intervention of any kind: that it be reversible, and that as long as the safety of the intervention is less than clear, "follow-up should also continue into subsequent generations."[28] Gregory Stock's argument favoring the artificial chromosome offsets this point. "Adverse effects that lie undetected for decades or generations are what most worry bioethicists, because by the time we notice them, many people might be effected"; however, as he a bit later goes on to aver, "[c]learly, this is moot with the nonheritable artificial chromosome I've described, but the argument is weak anyway. Because technology does not stand still, the more generations it takes for a problem to manifest itself, the more likely it is that medicine will have the expertise to deal with it."[29]

Stock assumes far too much in his advocacy of the artificial chromosome, as, technical matters aside, there is some question about the health dangers of humans' carrying an extra chromosome.[30] More categorically, the possibility of irrevocable damage appearing down the road is anything but "moot." Whether or not the damage might be significant would depend upon the nature and severity of the harm, matters eluding any knowledgeable assumption before the fact. Behind the tendentious nature of Stock's position there lies, however, another point that emerges where, dismissing what is moot, he refers to what is "weak" in the critics' position. The critics' concern for safety is much overwrought, as Stock sees things, since the more time it takes for a problem to appear, the greater the likelihood that science already will have come upon the medical means of dealing with it. This belief makes a monumental assumption. It assumes that irrevocable harm ought not enter our thinking because, whatever the problem might be, medical science is likely to have the expertise needed to treat it. Science not only can, but most likely *will*, possess knowledge of what might very well be both unforeseeable and unprecedented. What surfaces quite plainly is Stock's faith in science's possessing answers even where the experimental means of controlling causality goes wanting; indeed, even where the right questions pertinent to cause and effect may not yet have been asked. Such faith clearly expresses the belief in mastery subtlety abiding in the genetic engineers' widespread willingness to take risks. Belief in mastery is, indeed, the unstated faith behind the designers' willingness to step into the unknown with far more than a modicum of confidence in the science and technology guiding the way.

Belief in mastery runs, to be sure, throughout all of the thinking we have been discussing in this section. In the consideration of risk, the belief is subtle and assuages fear: How comforting it is to know that all science entails risk, as Venter tells us; that risk-taking is part of our being human, as Raam informs us; that our risk-taking explains the development of human civilizations beyond the African savanna, as Wade instructs us; and that science will be there for us with appropriately effective weapons should any nasty consequences appear down the road, as Stock assures us. Belief in human domination announces itself more boldly, of course, on the front and center stage of genetic manipulation where the master designers herald our approaching capacity to make nature over entirely to suit our own purposes. The protean nature of genetic material is today already widely evident in our ongoing genetic *enhancement* of plants and animals and, most remarkably perhaps, in our creation of chimeras that were hitherto precisely chimerical in their encasement in myth. And since nature is assumed not to place limits on our developing capabilities, the belief in domination ultimately points directly to the design of our species' genetic future. While addressing the possibilities of individual enhancement and family choice may be pleasing to the gene weavers, it does not take long for larger ambitions to arise in the design of species enhancement.

In contrast to the veiled presence of mastery in the acceptance of risk, the presumption of human domination of nature gains the spotlight where the idea of genetic enhancement speaks directly to species development. When the discussion turns to engineering human evolution, matters of safety and risk recede from view much like earthly things drop from sight for the astronaut whose flight takes his or her vision through the troposphere. What also recedes, or at least becomes unclear, is the role individual choice or family choice is supposed to play at the level of species development. This matter is fraught with political questions rooted in the issue of eugenics, which, though a critical issue, lies beyond the concern of this study. What concerns us, rather, is what genetic designers speak of when their thinking about enhancement leaves behind the domain of personal improvement and turns to humankind as a whole. In this larger domain genetic designers speak variously of *directed evolution*, *self-evolution* or *self-directed evolution*. These terms expressly identify instrumental adaptation with the belief that we can subject our own species' constitution to maker's knowledge in the interest of creating an improved version of what it means to be human.

Natural selection becomes anachronistic. It is a thing of the past, as the human material of organic change will for the first time become subject to self-conscious manipulation according to blueprints fashioned by maker's knowledge. As selection genetically passes from nature to human hands, design will replace randomness in our species' evolution. Favorable variations will no longer fall under nature's determination but be decided by whatever humanity favors as it contemplates its own future(s). This development is not necessarily inconsistent with Darwinian understanding, and Gary Cziko argues in his *Without Miracles: Universal Selection Theory and the Second Darwinian Revolution*, that our genetic control capabilities, or what he calls forms of "artificial evolution," arose randomly through natural selection with the emergence of adaptive complexity. Perhaps it is this element of randomness that keeps in Cziko's mind the darker side of future possibilities. As he writes, our newest scientific-technological revolution "brings with it the unprecedented ability to direct evolution itself.... Whether it will be positive or negative for the long-term survival of our and other species is, of course, the big question that only time and evolution itself can answer."[31] Cziko, unsurprisingly, does not tell us what evolution might mean apart from human design.

Cziko's critical concern tends to be exceptional, as genetic designers typically read the meaning of progress into self-conscious evolution. Even as cautious a thinker as Ronald Green can tell us that "[g]enetic science has opened our biology up to self-construction and directed evolution. We will certainly try to bring our biology under our control as we have done with so much of nature.... Emerging genetic technology permits us to replace the destructive and wasteful process of natural selection with intelligence and design."[32] Just as we can select what we favor, so, Naam tells us, our descendents may choose among designs departing from the species mold called human today. Humans "are not the end point of evolution — there is no such thing. We are just an intermediate step on one branch of the tree of life. But from this point on, we can choose the directions in which we grow and change.... We are, if we choose to be, the seed from which wondrous kinds of life can grow. We are the prospective parents of new and unimaginable creatures. We are the tiny metazoan from which a new Cambrian can spring."[33]

Whoever and/or whatever may follow upon the new Cambrian age, our instrumental adaptation will in this way be entirely plastic. Such open-endedness can exhilarate design engineers in the way genetic progress might be thought a menu consisting only of blanks to be filled in and ordered up as values change in time along with expanding design possibilities; or, in Rifkin's metaphor, people will have the "ultimate consumer playground," bringing about "the greatest shopping experience of all time."[34] Approaching the point of self-creation, we will come to make of ourselves whatever we wish to be, and it is difficult even to imagine what possibilities lie down the road. Silver's *Remaking Eden* is smartly entitled, and his epilogue to that work is telling in this same regard. There, in a futuristic scenario, Silver imagines that by the twenty-third century C.E. the enhanced who follow us will possess a "knowledge" and "power" far transcending our own understanding of these terms. From their tremendously advanced vantage point they will find themselves finally facing their creator with the major questions traditionally called metaphysical on their lips. "What do they see?" Silver asks. "Is it something that twentieth-century humans can't possibly fathom in their wildest imaginations? Or is it simply their *own* image in the mirror, as they reflect themselves back to the beginning of time...."[35] Presumably, the super-enhanced will answer all the old, outworn metaphysical questions since they themselves are their own maker.

Silver's thinking is unwittingly ironic since he devotes another entire work to criticism of the *spiritual* roots of the perspectives opposing genetic engineering. In what, after all, is his own futuristic scenario rooted if not a spiritual faith in genetic science and the human progress to be sprung from that science? Stock is much more explicit in this particular regard when he writes the following:

> [E]mbracing the challenges and goals of these transformative technologies is an act of extraordinary faith.... In offering ourselves as vessels for potential transformation into we know not what, we are submitting to the shaping hand of a process that dwarfs us individually. In secular terms, this is nothing special: we are merely accepting the possibilities of the advanced technologies we are creating. But from a spiritual perspective, the project of humanity's self-evolution is the ultimate embodiment of our science and ourselves as a cosmic instrument in our ongoing emergence.[36]

These are heady words indeed. Not all who do molecular biology promote genetic engineering. Nor do all of those promoting biotechnological intervention in the human genome necessarily subscribe to the perspective(s) we have been treating here in this section. Yet Stock's words do expressly identify a faith and spirit shared by a good number of the biological cognoscenti who seek to inspire and instruct the rest of us on the world-historical empowerment genetic science and technology are about to confer upon humankind. And they seek our support because when all is said and done, as they view the future, impossibilities will continue to be defined by people, not nature.

When belief in mastery comes clothed in biotechnological material, instrumental adaptation becomes one with genetic design and the control of human evolution. Interactional relations between human beings and their environment(s) are not altogether ignored as factors in adaptation; environmental influences are admitted in some consideration of personality formation, intelligence quotients, disease factors, etc. However, in this genetic understanding of instrumental adaptation, the overall significance of things lying outside genetics receives slight weight. For the genetic designers, genetics and human choice gain the spotlight; they are the controlling factors of organic change, leaving all else dimly in shadow. Exploring some of those shadows will be our task in the next sections of this chapter. We want to examine the refractory character of organic change on the ground floor of species adaptation, where novelty and contingency can and do elude human control in the most elementary manner. Our material of choice will be human disease, which obviously speaks substantively to some of the presumptive claims of the genetic masters. But before turning away from the genetic masters' vision of instrumental adaptation, let us see that vision once more:

> *Homo sapiens* has overcome the limitations of his origin. He controls the vast energies of the atomic nucleus, moves across his planet at speeds barely below escape velocity, and can escape when he so wills. He communicates with his fellows at the speed of light, extends the powers of his brain with those of the digital computer, and influences the numbers and genetic constitution of virtually all other living species. Now he can guide his own evolution. In him, Nature has reached beyond the hard regularities of physical phenomena. *Homo sapiens*, the creature of Nature, has transcended her. From a product of circumstances, he has risen to responsibility. At last he is Man. May he behave so!"[37]

This clarion call for instrumental adaptation à la genetic design—this vision of Man-the-Master—comes not from the work of any contemporary molecular biologist or science writer. This statement concludes the 1968 report on the biological sciences written and published by the National Academy of the Sciences. Considering this source, it is hard to imagine there being any better evidence of Lewontin's point that the manipulators' advantage over the observers lies in the fact that our culture already is given to the belief that the role of science is to subjugate nature. We have yet to see things clearly beyond the intoxicating legacy of Newtonian science, and in proposals for the genetic control of humanity's future, belief in human dominion persists as dangerous illusion. Exactly how dangerous this illusion may be is a matter lying inescapably beyond our ken. It may well be a question for storytelling; not in our time, but perhaps in the generation or two ahead.

Moving Targets

Fanfare of the highest order accompanied the public presentation of the Human Genome Project in June 2000. Albeit the celebration of a project only 85 percent complete at the time, President Clinton's White House cohosted the event with England's Prime Minister Tony Blair, who was hooked up to the White House via video satellite. Alongside Clinton were the Project's primary authors, Watson, Francis Collins, head of the National Human Genome Project Institute, and Venter. Together addressing the "most wondrous map ever produced by humankind," to employ the metaphor Clinton's speech writers had given him, these men and others present gave common voice to the hope and promise of our following the genetic directions this map might provide for the medical cure and elimination of human disease.[38] Hyping a project that had cost something in the neighborhood of $3 billion clearly had something to do with the show, but just as evident was the deeply held belief that our species had just turned an historic corner in its age-long battle with disease.

Given time and proper research money, genetics might finally bring to that battle the keys to ultimate success.

This wishful thinking directs attention immediately to genetically based diseases like Parkinson's and Huntington's, while more loosely referring to the genetic component of all diseases, whatever may be their source or nature. For some time now much thinking about disease has been undergoing a conceptual shift begun during the last third of the last century. Medical thinking about disease increasingly has come to link, and in many cases restrict, attention to genetic factors, to the point where "genetic disease" has become a major construct in the medical community. Arguing that this construct is now dominant, Evelyn Keller points out that the medical sciences welcomed this concept of genetic disease for a wide variety of institutional and economic reasons, and that while the concept is an "ideological expansion of molecular biology far beyond its technical success ... it was the concept of genetic disease that created the climate in which such a project [as the Human Genome Project] could appear both reasonable and desirable."[39] Whatever relative success medicine someday enjoys on the genetic front of this or that particular disease, approaching disease from a genetic angle certainly shifts attention away from issues involving social justice (how, where and why certain groups of people are subject to greater risk of disease) and social responsibility (who or what is contributing to conditions adversely affecting our health).

This positivistic tendency towards genetic reductionism inclines thinking away from the interactional manifold generating much human disease. At the same time, the microbial world of disease falls beyond the compass of such thought to a large extent, when, in point of fact, better appreciation of that microbial world and the problems it poses for medical science might, at the very least, help to establish some critical perspective on the likelihood of humanity's ever ridding itself of disease. Bacteria provide a good launching point for discussing what we are up against in that world, especially since the appearance of penicillin in the mid–twentieth century led medical science at that time to declare victory over bacterial disease without really knowing what it was facing.

The British bacteriologist Alexander Fleming discovered penicillin when by accident he observed the effect a *Penicillum* mold produced on a nearby bacterial colony. Extracting the material he would call penicillin from cultures of the mold, he proceeded to explore its deadly effect on various bacteria, the most important of these being the bacterium *Staphylococcus aureus*, which causes a host of skin infections that can lead to fatal septic disease. Fleming made the cover of *Time* magazine in the spring of 1944. The original "wonder drug" was at hand, or so it seemed with pathologist Howard Florey's widely successful promotion of penicillin's clinical successes. Yet right at the outset Fleming himself gave reason for caution in the public relations department as he observed resistance to penicillin emerging among surviving strains of bacteria.[40] He warned specifically how both overuse of the drug and small, insufficient doses would foster resistant strains that might easily pass from one person's body to the next with the migration of immune bacteria. Fleming further fretted about the onset of oral forms of the drug that could place its usage outside the jurisdiction of hospitals and thus magnify the potential for resistance. In point of fact, penicillin enjoyed almost a decade of free, over-the-counter dispensation before becoming a prescription drug in the mid–1950s. Why anyone would want to limit access to a wonder drug nevertheless became clear fairly quickly, as official reports coming out of city hospitals soon bore out Fleming's worst fears. Those reports concerned increasing bacterial resistance.

Had the potency of penicillin been restricted to *Staphylococci* bacteria, the discovery of this natural drug still would have been widely celebrated. Staph infections were the bane of hospitals everywhere, having long plagued hospital care, and the promise of penicillin to relieve pain and reduce mortality on that battlefront was immense. That promise soon became qualified, however, when reports of spreading staph resistance began to appear. One city hospital reported in 1946 that 14 percent of *Staphylococci* strains already had become resistant, and then four years later that same hospital reported a new figure of 59 percent.[41] And that was just the beginning. Confined at first to city hospitals where the sick typically gather in large concentrations, resistance spread outward throughout community hospitals in the 1960s and 1970s. Of course, the research behind treatment did not stand still during those decades. Broad-spectrum antibiotics like the tetracyclines proved to be effective over a wide range of microbial activity, and within the penicillin family ampicillin and semisynthetic methicillin were introduced to take up the slack caused by resistance to penicillin. Yet these solutions proved to be equally short-lived, as bacterial resistance caught up to them as well.

While Fleming's initial concern about bacterial resistance was entirely correct as far as it went, neither he nor anyone else at the time had any real idea of what they were up against. Resistance was thought to operate conventionally through natural selection, with perhaps a mutation or two thrown in. As susceptible microbes die, empty niches arise for the few resistant bacteria to occupy; and as this occupation spreads, resistant strains multiply through successive generations. Bacteria actually are much better than most other living things at this straightforward passing of adaptation from one generation to the next. But the adaptive playbook used by bacteria also includes an additional play, a lateral between contemporary species that can set up an altogether novel forward pass. It took almost two decades after the introduction of penicillin for full awareness of this lateral genetic play to crystallize among medical defenses. By then the contest was well underway, however, and this play had enabled many microbial pathogens to take the offensive and even the score against human ingenuity.

The organic change under discussion here caught no one's attention until dysentery began running through a large number of Japanese hospital patients in 1959.[42] The culprit in this case was the bacterium *Shigella dysenteriae*, which shockingly proved resistant to tetracycline, sulfonamide, chloroamphenicol, and streptomycin. Multiple resistance of this sort raised questions going beyond conventional mutation or simple variational selection. Japanese scientists realized that mutation rates made it astronomically improbable that the *Shigella* could have so rapidly evolved an enzyme capable of disarming all four of those drugs. What the scientists then discovered in the patients' infected diarrhea was an identical combination of resistances, the very same resistance profile, in a different bacterium, the *Escherichia coli* (*E. coli*), which is the normal intestinal flora in each of us. There was but one inference to be drawn from the fact that, in the same stool, two radically different species of bacteria shared the same extraordinary capacity for multiple resistance to the same drugs. Resistance had been genetically transferred from one species to another, which meant that DNA was transmissible apart from chromosomal inheritance. This discovery found quick confirmation in both Europe and United States, and, in Stuart Levy's words, it "opened the eyes of microbiologists and medical scientists to a breadth of gene spread never before imagined. Transfer of resistance genes could occur among bacterial species more genetically and evolutionarily distant than a horse is from a cow"; he continues, "Though not fully realized at the time, these findings predicted the widespread nature of antibiotic resistance we face today."[43]

Adaptation has gone very well for the bacterial pathogens targeted by our biomedical assaults. Without attempting any long, detailed story of staph resistance, suffice it to say that the bacteria were anything but cowed by the continuing determination of biomedical science. Upon the heels of penicillin's growing ineffectiveness the medical community in the early 1960s introduced methicillin, but soon there arose strains of methicillin-resistant *Staphylococcus aureus*, or MRSA, which also quickly proved resistant to flouroquinolones, a family of completely synthetic drugs developed in the 1960s.[44] Perceiving MRSA as a major threat to the treatment of staph infections, the medical community in the 1980s then turned to vancomycin, a relatively expensive antibiotic, as the preferred antibiotic. For a time vancomycin proved highly effective against all strains of staph, but, unsurprisingly, vancomycin-resistant MRSA showed up in the 1990s.[45] Near the end of the millennium two other new antibiotics appeared. One was structurally related to antibiotics that already were ineffective, so its effectiveness predictably proved short; but the other, linezolid, or zyvox, was the first structurally new antibiotic to be introduced since the early 1970s. Structural innovations generally are preferred as the best bets in antibiotics because their operational mechanisms are assumed to be unknown in the microbial world. MRSA proved to be an especially fast learner, however, for within two years resistant strains already were evident.[46] At the present time, barely a decade into the new millennium, there exist staph strains called Extensively Resistant Staph, or EMRSA, that are immune to every known antibiotic.

EMRSA is on the loose, and the stakes are large. In 2006 the National Institute of Allergy and Infectious Diseases (NIAID) reported startling statistics provided by the Centers for Disease Control and Prevention (CDC) headquartered in Atlanta, Georgia.[47] Each year in the United States alone almost two million hospital patients acquire infections while in the hospital. Over 70 percent of the bacteria causing these infections are resistant to one or more of the antibiotics used in treatment, and of the nearly two million patients annually infected, 90,000, or about 5 percent, die. It is shocking enough to know that this year 90,000 Americans entering the hospital will not leave it because they will fall prey to predatory microbes simply awaiting their arrival at the health site. More disturbing yet, however, is the frightening fact that this figure is seven times the number of such hospital-acquired deaths recorded in 1992.

Staph is but one of the infectious bacterial diseases alarming the medical community today. Gonorrhea, pneumococcal pneumonia, and malaria often are cited in the relevant literature, as is tuberculosis, which is especially worrisome given the number of human beings carrying this disease. The World Health Organization (WHO) estimates that more than one-third of the world's population carry one strain or another of tuberculosis, with five to ten percent of the HIV-free population developing the disease at some point in their lives.[48] The vast majority of people do not succumb because the human immune system normally protects us from TB. When some immune-suppressing condition such as HIV enters the picture, as it does widely in Africa, however, the five to ten percent becomes magnified horrifically. Deaths from TB numbered about 1.6 million globally in 2005, and while per capita mortality has declined, population increases have led to increasing deaths from TB.[49] What we confront in TB is an enemy all too ready to form alliances with killers like HIV. And, like other bacterial pathogens, TB has developed resistant strains rendering treatment monumentally expensive as well as extremely problematic from a medical point of view.

Microbial resistance to antibiotics follows a familiar story line. The wonder drug targeting *Mycobacterium tuberculosis*, the bacterial agent causing TB, was streptomycin, which made a much ballyhooed appearance alongside penicillin in the 1940s. Streptomycin suffers the drawback of relatively high toxicity, but that concern became somewhat moot once resistant strains of *M. tuberculosis* appeared, which did not take long.[50] Other drugs then replaced the fallen phenom, with the powerful isoniazid and rifampin emerging as effective drugs of choice. These last two antibiotics became a front line of defense, but when *M. tuberculosis* showed resistance to them in the 1990s, Multiple Resistant TB, or MDR-TB, became the urgent concern of all researchers and caregivers engaging the disease. Second-line antibiotics following up first-line treatment seemed the next best thing to do medically, but soon the pathogen responded by showing resistance to all such treatment. At that point, Extensively Drug Resistant TB, or XDR-TB, became public enemy number one. And rightly so, for XDR-TB joins EMRSA in the exclusive club of novel microbial pathogens that are essentially resistant to all known antibiotics.

WHO declared TB a major threat to world health back in 1993 and since then has undertaken major campaigns to halt the spread of the disease. In 2006 that organization launched its second long-term plan. Calling MDR-TB and XDR-TB "a threat to the security and stability of global health," WHO Director-General Margaret Chan allocated $2.15 billion specifically to deal with these novel strains of the tubercle bacillus.[51] This dollar investment in research and diagnostics is high, but then the cost of treating these extreme strains of TB is itself vertiginous. Treating MDR-TB can cost 1,000 times more than the cost of treating standard TB, and in 2006 the reported cost of treating one particular XDR-TB patient in the United States was just under $500,000.[52] It is true that these two strains of tuberculosis remain rare in the United States, but by 2007 thirty-seven countries had reported cases of XDR-TB, and WHO estimates that each year there are somewhere between 25,000 and 30,000 new cases of this recalcitrant menace.[53]

Researchers continue searching for new antibiotics, especially those having some structural innovation, and there is hope that the widening use of genomics to gain greater insight into the enemy may occasion the creation of an antibiotic that can target the pathogen without indiscriminately attacking all susceptible strains in the scattershot manner of broad spectrum antibiotics. In speaking of this continuing struggle, NIAID director Dr. Anthony Fauci spoke tellingly in a National Public Radio interview he had with Steve Inskeep on NPR's "Morning Edition" in early June 2007. Hearing Fauci talk about staph and TB and about the "damper" pathogenic resistance generally has put on antimicrobial advances made over the decades, Inskeep asked him, "Can you imagine a time when we might have to accept much higher rates of infectious diseases and deaths, perhaps like we had a century ago?" Fauci replied as follows:

> No.... It's a threat. I will not submit that we have to accept that one day that's going to happen. *I have confidence that the research community will keep up*, but that's not going to happen automatically. It's going to be kind of an agreement between the fundamental research community and the developmental community and the pharmaceutical organizations. If we get asleep at the switch, then it's going to happen. But we should not allow that to happen [italics added].[54]

The contest will continue, and vigilance and ongoing commitment to research will hopefully be enough to *keep up*. Keeping up is our best bet with moving targets.

Fauci's sensibility is a long remove from the hope articulated by the German chemist Paul Ehrlich, who at the turn of the twentieth century spoke the dream of a "magic bullet"—a drug that might eradicate infectious bacteria without in any way injuring the patient. This dream seemed to materialize in the wonder drugs at mid–century. As Stuart Levy, one of the world's foremost experts on microbial resistance, tells us, "the discovery of penicillin took on mythic proportions. It was as if Prometheus had stolen fire from the gods. The applications of this wonder drug seemed almost limitless.... These distorted expectations became a part of the mystique surrounding penicillin and other antibiotics, which has been carried over even into present times."[55] Bacterial pathogens from the start have been dodging magic bullets and forcing Prometheus to keep up. These moving targets certainly have proved able to hold their own in the contest with medical science, and perhaps the wisest counsel in this ongoing struggle speaks not of aggressive offense, but of prudent defense; not of increasing antibiotic use, but of prudentially curtailing that use where use actually amounts to misuse.[56]

Misuse may generally be identified with non–therapeutic applications and with those therapeutic applications that cannot achieve any therapeutic effect. In this latter category are situations of overuse, which is most flagrant in middle-class American patients' demanding and their physicians' providing antibiotics for "whatever ails 'em," as well as situations of underuse, which occur when dosages are inadequate to the task either because of cost or because of limited availability. The largest and truly most egregious case of misuse of antibiotics, at least in the United States, arises, however, with the non–therapeutic use of antimicrobials by livestock producers. At issue here is not the treatment of livestock disease but the use of antibiotics to enhance animal growth or to offset the adverse health effects of poor sanitation and overcrowding. Some time ago European countries legally prohibited such indiscriminate applications, but misuse continues to be widespread in the United States where, with related drugs, antibiotics applied to livestock amass to eight times the amount medically used by human beings. Not surprisingly, the Union of Concerned Scientists has made reduction of antibiotic use in animal agriculture the centerpiece of its antibiotic resistance project.

What makes misuse different from proper use is that it unnecessarily and unwittingly aids and abets the multiplication of resistant strains. Since destroying strains of bacteria susceptible to antibiotics creates empty niches that are filled by resistant strains, resistant bacteria flourish where drugs eliminate their competition. Curtailment of misuse thus aims to establish a defensive Darwinian alliance with those susceptible bacteria that are in competition with already resistant strains. This strategy makes good sense but guarantees nothing in the way of victory. As Levy writes, although "we know that antibiotic usage has contributed to the selection of resistant genes, we do not understand what causes the sustained, relatively high level of resistances in areas and countries where the antibiotic has been stopped."[57] It appears that the contest will go on even under conditions meant to favor our bacterial allies.

As if any further bad news were needed from the front line, recent bacterialogical research reports a discovery that argues all the more for the elimination of any and all misuse. What scientists are calling an *antibiotic resistome* are bacteria that sustain themselves entirely on antibiotics. Subsisting on antibiotics, these bacteria could be making multiple resistance genes available to those bacterial strains standing nearby, just awaiting a lateral handoff. Scientists recently reporting their investigation of antibiotic resistomes in *Science* concluded their overview of the resistance camp, as follows:

> In addition to the finding that bacteria subsisting on natural and synthetic antibiotics are widely distributed in the environment, these results highlight an unrecognized reservoir of multiple-resistant machinery. Bacteria subsisting on antibiotics are phylogenetically diverse and include many organisms closely related to clinically relevant pathogens. It is thus possible that pathogens could obtain antibiotic-resistant genes from environmentally distributed super resistant microbes subsisting on antibiotics.[58]

We need not imagine the worst-case consequences suggested here to appreciate the resourcefulness of these microscopic, single-celled organisms. In numbers, bacteria are the dominant form of life on earth, and without them human life would not exist. Belief that we will conquer those bacteria that in any way threaten us stretches credulity, to say the very least. The bacteria themselves certainly are not backing down, and their means of survival seem, as often as not, to elude our shock-and-awe tactics quite effectively. Knowledge of what we are up against in the bacterial world is what led the World Health Organization at the turn of the millennium to identify antibiotic resistance as one of the three major threats to public health in the twenty-first century.

Surprise Attacks

Bacterial resistance illustrates in plain elementary fashion how life forms can adapt themselves to instruments of applied science, and, in so doing, not only frustrate our aims, but, moreover, interactionally strengthen themselves. Turning attention to viral diseases leads us to an altogether different set of problems having to do with the particular nature of viruses and the unique character of the organic change viral microbes can undergo. Organic change may operate on the viral front adaptively in ways similar to what takes place in bacterial resistance, but in its most dramatic appearance, viral change announces itself in emergent diseases, which define themselves in their absolute novelty.

Viruses have posed terribly difficult if not intractable problems for medical science right from the start. No wonder drug has been promoted in the face of viral disease because the very nature of viruses runs counter to the notion of a magic bullet. A virus is miniscule stuff alongside a bacterium, being anywhere from ten to 1,000 times smaller. Not all viruses even are visible through our most powerful microscopes. Genetic material illuminates this difference even more: whereas the tiniest bacteria possess between 5,000 and 10,000 genes, the viruses causing measles, Ebola, polio, and HIV have fewer than ten genes each.[59] More significant still, while bacteria are unicellular microbes, viruses constitute a third category of nature, certainly outside of plants and animals, and somewhere between the animate and the inanimate. Whereas bacteria are cellular, viruses are not, being "nothing more than a speck of genetic material" consisting of DNA or RNA with a protein molecular coating.[60] A virus' non–cellular nature at the same time makes it parasitically needy in a very special way: a virus must inhabit a host cell in order to do anything. Unable to replicate on their own and possessing a status somewhere between living and non–living things, viruses need to gain entry into foreign cells where they can take over existing functions in a kind of invasive occupation. Within a host cell a virus can replicate, utilize cellular material in various ways for its own purposes, and hide for as long as it wishes or needs—perhaps decades—before launching a major offensive. Once it does launch a major offensive, the virus then typically holds the cell hostage to counterattack by viral drugs. It is true that the absence of cell walls enables viruses to elude those drugs that specifically attack or take hold of a cell wall, but more generally viruses find protection within or "behind" their host cells, which would suffer full collateral damage in the case of a direct medical counteroffensive.

Given the severe limitations of antiviral drugs, science has had to employ battle strategy different from that used against bacteria. Vaccination is the name of that strategy, which is defensive through and through. Prevention rather than counterattack has been the watchword in this battle. Vaccines can lose their defensive potency, of course, and new strains of a virus can and do render existing vaccines ineffective. Many cases of long-term success, nevertheless, trumpet the appropriateness of this approach in general. Among those successes, the eradication of smallpox is itself perhaps the most significant achievement in medical history. But wherever success cannot be claimed, we find the same refractory character of organic nature. And viruses exhibit such character brazenly, in routine fashion as well as in dramatic surprises. Influenza, or the flu, offers as good an example of routine viral ingenuity as we are likely to find among the everyday viruses.

Influenza sounds terribly ominous whereas *flu* sounds old hat. Not uncommonly, the former word is coupled with "epidemic" or "pandemic," denoting the kind of plague that swept the world at the end of the Great War. *Flu*, on the other hand, domesticates the virus. Flu lies within the cycle of life's rhythms; it has its season each year, and each year we are told to get our flu shots. Over thirty thousand deaths annually are attributable to seasonal flu in the United States alone, the elderly being the most susceptible, but seasonal flu seems far from menacing to most of us. So things normally appear, at least to the general public. Lying behind those flu shots, however, is a vigilant monitoring network always on the alert and ever ready at the alarm.[61] WHO annually prescribes the contents of the flu vaccine in accordance with information and recommendations supplied by the CDC, which is responsible for monitoring whatever strains of flu get reported throughout the year. The CDC annually assesses anywhere from 10,000 to 15,000 reported strains of flu and determines for each global region which three strains need to be addressed in the particular vaccine likely to be most effective in that specific geographical area. Large numbers of lives regionally depend upon whether or not the CDC "gets it right," though getting it right cannot insure against the rapid ongoing mutation of those three strains targeted in the vaccine. During the 2007 flu season in the United States, authorities warned that two of the three viruses targeted already were mutating while the increasingly ineffective vaccine was being supplied. Not surprisingly, hospitals everywhere that winter were visited by large numbers of patients. The flu that most of us live with, is, in fact, serious business.

The stakes are incalculably higher in the CDC's regular monitoring of reportedly novel or unusual flu strains that might trigger a pandemic. There is no rest for those so engaged, for they know the nefarious character of influenza: they are on the lookout for an emergent flu having no precedent. There are three types of influenza virus.[62] Flu shots pay no attention to the Type C virus, which causes only minor respiratory illness and is believed to pose no threat of an epidemic. Included in the annual vaccines are types A and B, as these are the culprits of human suffering. Type B viruses are single-minded, infecting no species other than our own and cleverly practicing *antigenic drift*, by which they alter their antigens so as to disguise themselves to elude the antibodies our own bodies send out to defeat them. Drift is gradual but regular, and it is this common viral practice that requires the annual development of new vaccines. But there is another viral practice, *antigenic shift*, which occurs not gradually but all at once. This practice belongs not to Type B viruses but to Type A viruses. Speaking of antigenic shift, the CDC has this to say: "Shift results in a new influenza A subtype. When shift happens, most people have little or no protection against the new virus. While influenza viruses are changing by antigenic drift all the time, antigenic shift happens only occasionally."[63] "Occasionally" is the crucially fortunate term here since Type A viruses are not single-minded and affect various species of animal. Viral change in another animal species can lead to zoonosis, or the passing of a new viral subtype to humans. It is then that organic change can be menacing beyond belief.

All flu seems to have sprung from avian origins, which probably explains why many flu viruses are so adept at recombining and reassorting their genetic material, swapping or sharing new DNA across species while creating new viruses. Non-human animal species are far more susceptible than we to Type A viruses' swapping genes or gene segments. The relative rarity of a deal being made directly between a human virus and another animal's virus is, nevertheless, no cause for our dropping our guard. The menacing possibility most often cited, as a matter of fact, involves a third-party mixer brokering such a deal. The CDC is worth listening to at some length on this point[64]:

> Pigs can be infected with both human and avian influenza viruses in addition to swine influenza viruses.... Because pigs are susceptible to avian, human and swine influenza viruses, they potentially may be infected with influenza from different species (e.g., ducks and humans) at the same time. If this happens, it is possible for the genes of these viruses to mix and create a new virus. For example, if a pig were infected with a human influenza virus and an avian influenza virus at the same time, the viruses could mix (reassort) and produce a new virus that had most of the genes from the human virus....The resulting new virus would likely be able to infect humans and spread from person to person, but it would have surface proteins ... not previously seen in influenza viruses that infect humans.... If this new virus causes illness in people and can be transmitted easily from person to person, an influenza pandemic can occur.[64]

Reassortment of this kind appears to have underlain the emergent virus that killed tens of millions of people at the end of the Great War. Called the Spanish flu because Spain led the initial reporting of pandemic disease, no one knows for sure where this catastrophe originated. One widely accepted belief, the "Haskell hypothesis," holds that the 1918 plague actually sprang from farm animals in Haskell County, Kansas, where military men picked it up and then spread it.[65] A leading British virologist argues, to the contrary, that the pandemic originated in China. Jeffrey Taubenberger, an American immunologist, however, accepts neither of these positions.[66] Such disagreement itself evidences the difficulty often obtaining in efforts to track down the causality of organic events that are novel and historically embedded in interactional lifetime. Whatever the Spanish flu's exact origin or pathways of early contamination, the virulence and transmissibility of that novel disease had fatal consequences far exceeding the battlefield mortality of the Great War. Emergent influenzas can, of course, be devastating even on a lesser scale. A new flu virus ordinarily causes symptoms to appear among 15 percent to 40 percent of an exposed population, and even a virus fatal in just over 1 percent of those exposed could result in deaths between on .5 million and 1.3 million in the United States alone.[67]

Today the general consensus in medical science is that a pandemic influenza will soon be upon us. Most virologists believe that before too long some genetic shift or other among viruses will strike us mortally. Given the historical record, we are "due": in 1919, the Spanish flu; in 1957, the Asian flu, which that year killed 70,000 people in the United States alone; then, in 1968, the Hong Kong flu, which over the next three and a half decades would claim the lives of over 400,000 victims in the United States. The question now is what's next? Maurice Hilleman, the world's leading vaccine expert until his death in 2005, believed that pandemic flues actually operate on a calculable cycle, but he was in a minority.[68] The H1N1 virus, better known as swine flu, caused a serious scare in 2009. It was a novel virus arising from some eventful antigenic shift, and bore the markings of a global threat. But, increases in pediatric deaths notwithstanding, the virus fortunately did not attain anything nearing horrific proportions. A safe vaccine quickly became available to hundreds of millions, and, more critically, the virus itself proved to be far less a threat than had been feared. The CDC reported that, two weeks into 2010, total H1N1 deaths in the United States had reached just a little more than 11,500, and Paul Palese, a well-known flu virologist at New York City's Mt. Sinai School of Medicine, has felt it safe to judge H1N1 a "mellow virus."[69] The same cannot be said of the H5N1 virus, more commonly known as bird flu, which kept the World Health Organization ever alert and ill at ease throughout much of the new millennium's first decade.

H5N1 is a deadly flu virus, having a mortality rate in the vicinity of 50 percent for humans. The key worry from the start has been that some further reassortment of the flu's genetic material could eventuate a strain that is transmissible from human to human. Small wonder, then, that the world health community became terribly alarmed with the spread of the virus, which by 2005 had reached Southeast Asia, India, Southwest Asia, Turkey, and parts of Europe and Africa. Experts worried that increasing interaction between infected bird populations and human communities might effect the dreaded organic change, and when in 2005 tests run in Vietnam reported a strain showing real human-to-human possibility, WHO contemplated raising the alarm status it had given the flu to just short of pandemic proportions.[70] WHO restrained itself, and Vietnam proved a false alarm. Their having an itchy finger on the alarm trigger, nevertheless, is understandable.

Caution on the trigger had not been the case in the mid–1970s when an influenza scare coming out of Fort Dix, New Jersey, led the CDC to the federal government's door with a proposal to inoculate 200 million Americans in fear of an emerging pandemic.[71] Very quickly President Ford went before the nation, and five million people received vaccination while the swine flu never left the confines of Fort Dix. Over the next fifteen years, in the wake of death, illness, and paralysis caused by an insufficiently tested vaccine, thousands of Americans were granted $85 million in taxpayers' money for their unnecessary suffering. Those making policy judgments at WHO in 2005 were doubtlessly aware of this story. Meanwhile, the bird flu wants continued watching. Virologists do not fully agree that H5 viruses like H5N1 even are transmissible from human to human,[72] and though by early 2010 the bird flu had made no headway among humans, some uncertainty continues, and eyes remain closely watchful.

Medical science generally works with novel influenza viruses in this twilight zone of uncertainty where the future is not clearly visible in the gap between what we know and what we do not know. Such uncertainty may be hard to live with, but it is quite impossible to avoid. Flu viruses in any case are always on our monitor screens. We watch the flu family very closely because it is generally troublesome, and, moreover, because we know that every so often it produces an exceedingly bad seed. In other cases of novelty we may have no such previous awareness. Outside the flu family, altogether novel diseases can spring in places and from sources that are ordinarily benign, unknown, or at least unwatched until they give rise to something wicked. In such cases, the outbreak of disease truly comes as a surprise attack. One such case arose in 2002 and for a brief time had medical scientists both baffled and terribly frightened. This particular virus very quickly became known as Severe Acute Respiratory Syndrome, or SARS.

Like a number of other emergent viruses, SARS began in Asia where huge numbers of humans live with large numbers of other animal species in terribly dense demographic concentrations. The birthplace of SARS was Guangdong, a province in southern China, but we know nothing precise about the virus' parentage. Indeed, we really know very little at all about SARS. Everything relating to the outbreak and spread of the disease happened quickly, or so it seemed. Moreover, since SARS went into hiding in 2004, which is only one possible explanation for its disappearance, there has been little to learn about it apart from the original data gathered at the time, some of which is itself unclear. Perhaps the two most telling things about SARS are the very speed of unfolding events and the uncertainty surrounding its disappearance.

Chinese authorities did not inform the outside world of SARS until pressed to do so by WHO officials who had gotten wind of loosely circulating reports of a number of strange pneumonia deaths in China.[73] In early February 2003 China's official news agency admitted that there had been a few hundred cases of unusual viral pneumonia dating back to the late fall of 2002. The Chinese report was reassuring nonetheless: only five of the hospitalized patients had died, and the situation was under control. Nothing could have been further from the truth than their claim of having things under control. Later that same month one of the doctors who had attended stricken patients was himself suffering symptoms of the disease when he came into contact with a number of individuals in Hong Kong. Three weeks later WHO proclaimed SARS a worldwide threat. That Hong Kong contact had directly exported SARS to Vietnam and Canada, and the infection was well on its spread from there. In mid–May 2003 WHO reported SARS deaths in 28 countries. Then, ten weeks later and only four months after having declared the virus a worldwide threat, WHO announced the end of the global outbreak.

After having sickened 8,098 people, 774 of whom died, SARS seemed to have run its course.[74] It had been a surprise hit-and-run attack during which everything moved very quickly. SARS symptoms appeared after a relatively brief incubation period, and some of those infected immediately carried the disease with them aboard transoceanic jets, going from one continent to another. Researchers and medical communities worldwide advanced immediately with unprecedented speed in developing knowledge and defense work against the disease. And suddenly, several reported cases later notwithstanding, SARS simply vanished.

Identifying the virus had been critical, and researchers all over the globe directly had gone to work on the question. After German and Hong Kong investigators initially judged the virus wrongly, the right answer came within ten days of WHO's having rung the global alarm. SARS turned out to be one of the coronaviruses, which was extremely threatening news at the time inasmuch as coronaviruses, which had always been judged relatively benign on the whole, are the second most widespread cause of the common cold. Then, in mid–April, the British Columbia Cancer Agency announced their astounding microbiological feat of having sequenced the entire genome of a Toronto patient's virus. First identifying the virus and then mapping its genome with such celerity was a striking lesson in what can be accomplished in the best scenario, when global gears are meshed and focused. The bad report coming with this otherwise positive news was that SARS is completely unique; it is unlike any human or animal coronavirus. And this point spotlighted the second piece of bad news, which concerned our ongoing ignorance of the novel disease's origin. "The SARS microbe provides disturbing evidence that even when microbiologists have sequenced a genome, that alone may not tell them whether a pathogen jumped from animals, or if a previously unknown human strain simply mutated or picked up genes from other organisms. This information is vital in knowing how best to develop a drug to combat a disease."[75]

A good deal of guesswork surrounds SARS, and understandably so. At the time of the outbreak Chinese officials in Guandong Province carried out mass executions of civets, a wild-game animal, and current thinking is that animal markets caging wild game in Guandong may well have been this novel virus' point of origin.[76] The issue of drug treatment remains uncertain as well. An international team responding to WHO's request for a systematic analysis of treatment studies and in vitro investigations reported the following from its work: "From the published studies, it is not possible to say whether any of the treatments used against SARS were effective. No cases of SARS have been reported since 2004 but it is always possible that the same or a similar virus might cause outbreaks in the future. It is disappointing that none of the research on SARS is likely to be useful in helping to decide on the best treatments to use in such an outbreak."[77] Whether or not another SARS outbreak may happen is, of course, part of the larger uncertainty surrounding its sudden disappearance. Experts offer various explanations of SARS' vanishing act.[78] Two guesses lie on the sunny side of speculation. One guess is that once SARS jumped to humans, if that is what it did, it quickly evolved beyond human transmissibility; its communicability died with the last of those infected. The second encouraging explanation of SARS' disappearance is that SARS' evolution lessened its virulence to the point where now its symptoms can be mistaken for those of the common cold. Speaking to the darker side of speculative possibility, however, is our knowledge of how viruses have hideouts among all kinds of animal species, in which they can lie low for long periods of time. This particular explanation carries its own uncertainty inasmuch as there is no understanding of why SARS might have gone into hiding. Yet other viruses are known to go into hiding inexplicably, so this possibility is very real. In any event, Dr. Umesh Parashar, speaking as the lead epidemiologist on the CDC's task force on SARS, summed up the entire matter when he noted that the "'possibility for a resurgence remains.... It's something we have to watch for. I don't think anybody can put a frame on when we'll be out of the woods.'"[79]

Regardless of the issue of recurrence, SARS is an instructive case of a novel disease striking quickly and spreading rapidly. Emerging from a family of viruses that normally had been associated with relatively mild respiratory ailments, SARS' very novelty has blackened the name of coronavirus. In this case a very bad seed sprang from a relatively good family that had not been under close community watch because nothing in the previous record called any special attention to the normal goings-on of family members. Bad fortune struck initially in the Chinese authorities' failure to act with more alacrity and circumspection respecting quarantine and international notification. Once out, of course, the disease's rapid spread simply followed the normal pathways of global contact and exchange. As Levy and Fischetti tell us, the "incubation period for most infectious diseases is days, and it takes only 36 to 72 hours to circumnavigate the globe, so most new diseases can arrive before anyone knows about them."[80] Jet travel can and does turn local events into international affairs overnight. In the case of SARS, good fortune lay in the fact that its infectious stage does not correspond to the incubation period but arrives only with the onset of symptoms of illness. At that point, sickbeds rather than airplane seats usually are being sought.

Novel viruses are typically viral in their disdain for drug treatments, but what sets them apart is the challenge they pose for the development of vaccines. Successful vaccines are difficult enough to develop even for viruses whose backgrounds are well known, and the novelty of an emergent disease making a surprise attack magnifies all such difficulties. In the case of SARS the most optimistic guess was that a successful vaccine at best would require at least a year or better to develop and test for safety. Protection against the disease would take considerable time even if the first vaccine proved more than temporarily effective. A vaccine's long-term effectiveness in the face of possible mutation would be another, later question. More immediately, had SARS wanted to wage a protracted war rather than skirmish a few times, however long it would have taken to develop an effective vaccine might have allowed the virus sufficient time to have caused horrific casualties. It is fortunate that, at least within our experience so far, SARS proved to be merely a skirmisher. The same cannot be said about a number of other emergent viruses that have arrived on the scene in recent time.

Most infamous among the novel viruses currently plaguing humanity is, of course, HIV, and plague surely seems the apposite verb in this case. Deaths to 2008 totaled in the neighborhood of 25 million; there are today an estimated 30–35 million carriers; the effectiveness of the drug cocktail in use is of uncertain duration; and while the pandemic has been with us for almost three decades now, it will perhaps still be a number of years before an effective vaccine will be developed and proved safe. Inestimable are the efforts of tens of thousands of researchers whose careers have been shaped by warring with this disease, and most of the handful of relatively successful antiviral weapons we now possess are products of those efforts. A few principal characteristics make this foe especially intractable, and while these traits may be fairly well known, they are worth some discussion here.

Vaccines for smallpox and polio have enjoyed tremendous success in part because those particular viruses possess very slow mutation rates. The rate of organic change with HIV is an altogether different matter.[81] Being a retrovirus, this RNA virus invades a cell and uses it to produce its own viral DNA, which it then plugs into the genetic material of the host. Its multiplication entails opportunities for relatively rapid mutation, and mutation may threaten the long-term effectiveness of current drug treatments. If limited longevity of drug treatments is assumed, hopefully for argumentative purposes alone, HIV's mortality rate almost is 100 percent for those contracting the virus. Natural immunity to this virus favors but very few of the human population. Transmissibility is therefore of crucial importance. Apart from a fetus's or a child's being infected by the mother, HIV's infectious paths are subject entirely to choices that individual human beings make about how they will live and what they will or will not do to protect themselves. Exposure fundamentally is therefore either a matter of ignorance or of poor judgment at this point, and as education about HIV continues to expand, one would expect poor judgment to gain an increasingly larger share in responsibility for the transmission of the virus. In the beginning, however, transmission held hands with total ignorance.

HIV's extended dormancy was profoundly significant to the ignorance accompanying the early spread of the virus. Like other lentiviruses, which are characterized by the long period of time separating the symptomatic appearance of illness from the point of infection, HIV has unlimited patience. It can hide in a child's cells until it is ready to mount an attack on the grown man's immune system, at that point signaling the onset of acquired immune deficiency syndrome, or AIDS. This virus, indeed, chooses to lie low upon entry, as if preferring ambush to outright attack. That strategy proved most successful to the initial spread of HIV—when HIV was in its hideout and altogether unknown as a plague. We can appreciate this point by considering the story of HIV's early years, which we can stitch together quickly with some basic facts, including what may pass for facts today upon a scientific consensus.[82]

At this point there are three known instances of HIV dating before the 1980s, the earliest case being the death of a man in the Democratic Republic of the Congo in 1959. Of course, these cases of HIV appeared as such only in the after-light of the pandemic's appearance towards the end of the century. Recognition of an emergent plague followed upon the outbreak of AIDS in patients in Los Angeles, San Francisco, and New York City in 1981, and thus began the long war with HIV that continues today. Five years into the struggle the viral strains of HIV-1 and HIV-2 had been identified, and WHO stated that at the end of 1986 over thirty-eight thousand AIDS cases had been reported from 85 different countries.[83] Tacking between crackpot religious proclamations of divine punishment and the predictable conspiracy theories, serious scientific investigation of HIV has required two to three decades to piece together a plausible narrative explanation of where HIV began and how it may have spread. One matter in that investigation gaining early publicity was the neatly oversimplified story of "Patient Zero," a term referring to some initial point for the plague's dissemination. Patient Zero was Gaetan Dugas, a gay airline steward whose own promiscuity and travel opportunities combined to spread the infection among untold numbers of the twenty-five hundred men with whom Dugas estimated he had had sex between 1973 and 1983. The scourge of San Francisco's bathhouses, Dugas certainly was the key to that city's gay community's losing an immense portion of its population. He was, however, but one conduit coexisting among many others for the proliferation of the virus.

It is believed that by 1980 HIV had already infected perhaps as many as a few hundred thousand people on five different continents.[84] It is not difficult to understand how the virus could have spread exponentially given the increase in air travel, the prevalence of needle use among drug addicts, and loosened attitudes towards sex, to name but a few prominent factors. HIV-1 certainly took advantage of such ready opportunity, as this viral type, which is the more transmissible and virulent of the two, is the one most commonly found outside of Africa. HIV-2, the type predominating in Africa, could follow its own more local paths of transmission within that continent. In any case, there was no real doubt that the virus began in Africa, and research focused on various strains of the simian immunodeficiency virus (SIV) found in different monkey subspecies in central west Africa. The source of HIV-2 was directly traceable to one monkey subspecies that is commonly eaten or kept as a pet in that region of Africa. Locating the source of HIV-1 was much more difficult, however, as SIV viral strains found in monkeys were not quite right genetically.

Yet monkeys did prove to be the original source of the virus without being the zoonotic agent. Zoonosis came by way of chimpanzees, or at least one particular subspecies of chimpanzee identified at the very end of the last century.[85] Follow-up experimental research then concluded in 2003 that this subspecies of chimp had recombined SIV strains from two different kinds of monkeys it had eaten, becoming a kind of genetic mixer for a novel SIV disease.[86] Having genetically created a new simian disease, this particular chimpanzee thus became the source of HIV-1 once human beings took the new virus into their own bodies. Exactly how that interaction may have occurred is a subject of educated guesswork, most contingencies turning on the idea of African hunters' having come into contact with chimps' blood in one way or another, at some time or another, during the first half of the twentieth century. So goes in extremely rudimentary form the presently accepted narrative explanation of HIV's causal beginnings. The story takes on full meaning, of course, only with the surprise attack many decades later.

The story of this plague is altogether horrific. Yet humanity thus far has been fortunate with HIV in a way we can understand imaginatively. Consider the initial period of the pandemic's spread through the time of our complete ignorance — the ten years between, say, 1971–73 and 1981–83 — and let us change the narrative possibilities simply by adding one single element. Let us add one other means of transmission of HIV — the mosquito as a vector. Gaeton Dugas now becomes the name of each and every mosquito on earth that might draw blood from an infected person. Each mosquito bites one human after another, and this spreading of the virus among humans also rapidly multiplies the population of vectors since opportunities increase for mosquitoes to pick up the virus from human blood. Given exponential spiraling of HIV in the interaction of the feeders and the fed upon, how many winged insects named Gaeton Dugas would it have taken over the course of the decade to reach by such proliferation the vast range of humanity possessing no natural immunity against HIV? Should the decade of ignorance have been insufficient, another fifteen years of pre–cocktail time would have been available, of course. Nothing here should tax the imagination. Nor does the subjunctive mood of the scenario restrict our point to imagining a different past. Consider the future possibility medical journalist Pete Moore notes:

> Few diseases have highlighted the interaction of human behaviour, health, and infection, as has AIDS. Thankfully, it has never mutated into a form that can transmit by any other means than within body fluids. Let's hope it never finds a way of getting into the lungs and then moving on in aerosols as an infected person coughs. But the longer the pandemic goes on, the more chances HIV has of learning new tricks. It's in everybody's interest to stamp out this blight before it gets any worse.[86]

Hardly a comforting thought, this. HIV's initial surprise attack may be long behind us, but the element of continuing surprise remains in the protean nature of this viral scourge.

There are other frightening tales out there available for the telling: new hemorrhagic fever viruses like Ebola, West Nile virus, and many others. At the same time, just as known and established threats may be genetically altering their nature and varying their means of attack right before our very eyes, so, too, unforeseeable assaults by currently unknown novel enemies will most likely be forthcoming, if not tomorrow, then perhaps the day after. But we intend no speculation on such matters. We need, indeed, to be careful lest ultimately we miss the wood for the trees. In the end, the argument really is not about our won-loss record with diseases or whether or not a vaccine or treatment is attainable in any particular case. The point, rather, is that we are interactionally enmeshed in a vast and deep life-world in which organic change is ceaseless and threats to our well-being are inestimable. When the CDC and WHO confer on the frightening mutability of influenza, or when Fauci speaks of needing to keep pace with bacterial resistance, we witness the essential reality of our entrenched struggle to protect ourselves. Medical scientists and researchers need to be as proactive as possible, but as often as not they find themselves in reactive roles: trying to keep up, to counter, to fend off some biological phenomenon that is one or more steps ahead by virtue of its organic initiative and elusive causality. When and where our ingenuity and advancing knowledge find an advantage, genetic or other, we gain victory. Smallpox is an excellent example. Victory itself may prove quite temporary, however, as we see in the seasonality of flu shots and in the records of various antibiotics. Experimental research is most critical at the disease points of this dynamic interactional complex of organic change, but organic change really does pose insuperable problems to those who dream of eliminating disease by means of either genetics or some magic bullet. Life phenomena exhibit a mercurial nature that eludes our manipulative grasp and turns us into observers even as we continue to manipulate. Indeed, as emergent diseases make most clear in their novelty and surprise appearance, we become storytellers as nature confronts us with phenomena — with effects— we have never seen, via causalities unbeknownst to us at the time.

This same possibility of dire post–dictive knowledge also inheres in the risks taken with any proposal to control our own evolution by genetic design. Any pleiotropic or polygenic effects eluding experimental grasp in the present would themselves be emergent in the future, perhaps less likely as disease than as some abnormality of greater or lesser consequence. Since our knowledge respecting causality in such cases is largely uncertain, we cannot assess dangers in any reasonably meaningful sense. Delayed effects would be given as such only after organic change had lain in wait, as it were, through lifetime, as if wanting to make some dramatically sudden appearance. To deny our own organic nature this reality or our genes this possibility, or to assume that such emergent eventualities themselves would likely be of little consequence, given our progressive instrumental capabilities, is to live the illusion of mastery and to dream the dream of dominion — as if it is people's choices alone that might limit our knowledge and restrict our capabilities, but certainly not nature. Such hubris feeds upon illusion, for if one thing is certain, it is that there are more things in nature's possibilities and in our genes than are dreamt of in our genetic designs.

In the empirical cases we have addressed thus far, it is easy to spot the temporal domain of causality's appearance and the narrative form of our understanding. In the next chapter we will highlight life-time and narrative explanation more directly as we spotlight contingency and complexity in the interactional elements lying behind organic change and environmental change in several of the case studies ecologists and environmentalists have used in their attempts to develop a critically informed public consciousness of our endangered plight on this planet Earth. Our choice of case material is fully in keeping with the larger concerns of those critical ecologists, but our use of their material will focus solely on the limitations of instrumental knowledge. Problems of prediction and uncertainty will occupy our attention as we further criticize belief in mastery from within the organic manifold of nature's ecology.

6

Ecological Detective Stories

Less than a decade ago Nobel Laureate chemist Paul Crutzen wrote a very brief piece in *Nature* entitled "Geology of Mankind" in which he employed the term *Anthropocene* to designate a new geological epoch. Crutzen tells us there that the Anthropocene "could be said to have started in the latter part of the eighteenth century, when analyses of air trapped in polar ice showed the beginning of growing global concentrations of carbon dioxide and methane. This date also happens to coincide with James Watts' design of the steam engine in 1784."[1] Marking a shift in geological epochs, the Anthropocene takes over from the Holocene when, with the West's materialization of the project of mastery in the industrial revolution, humankind assumes prominence in the processes of planetary change. The idea of the Anthropocene has gained increasing currency with its emphasis on anthropogenic causality, especially in view of climate change, which may be this epoch's signature development at this point. In any case, humankind's instrumental intrusion into nature has reached the point where it is the interactional axis around which turns this planet's future.

We have highlighted our interaction with nature in critical discussion of both the quantum revolution and the phenomena of antibiotic resistance and microbial change. The interactional reality illuminated by the idea of the Anthropocene is itself, however, far more complicated by several orders of magnitude than anything we have encountered thus far. What comes to light here are extremely complex ecological situations in which environmental change and organic developments operate dynamically all at once on various interactional levels. Genetic engineering in this context is especially problematic respecting the countless genetically modified organisms we continue to insert into the environment in the interest of improved agriculture, among other things. Numerous critics over the years have addressed the possibilities of such artificial organisms' wreaking environmental havoc of one kind or another, and whatever the danger may be in this or that particular case, each "introduction is tantamount to playing ecological roulette," to employ Rifkin's telling metaphor. As Rifkin more generally puts the point of such uncertainty, "There is no predictive ecology to help guide this journey [of artificial organisms] and likely never will be, as nature is far too alive, complex, and variable to ever be predictably modeled by scientists. We may, in the end, find ourselves lost and cast adrift in this artificial new world we're creating for ourselves in the Biotech Century."[2] Our concern in this chapter lies not with the genetic modification of organisms, however, but with the problems that ecologically complex interactions more generally pose for our knowledge. None in the scientific community make better storytellers than ecologists, even though some among them would have us believe that the interactional realities with which they deal should ultimately be subject to some kind of scientific control.

The study of ecology concerns itself with interactional complexity, and in turning our attention to various ecological case studies in this chapter, we want to deepen our critique of mastery by spotlighting the limitations of maker's knowledge vis-à-vis the kinds of challenge posed by complex and contingent interactional realities. Moving onto ecological terrain also will lead us ultimately to the thinking of planet managers who attack those challenges from an interactional perspective, claiming instrumental control capabilities different from those we have defined as maker's knowledge. Before examining the nature of their answer to those challenges, however, we need first to read through a collection of ecological detective stories.

An Interactional World

Rachel Carson's *Silent Spring* appeared in 1962, the child of four years' labor. *The New Yorker* already had serialized this work, but it was *Silent Spring* itself that single-handedly opened critical eyes to serious ecological problems and permanently placed environmental issues on governmental agendas, university curricula, and corporate lobbyists' lists. A rare achievement of impassioned scientific expertise, literary style, and political punch, this work has no equal in either its appeal or its consequentiality at the intersection of science and public affairs. Indeed, its author, who died just two years later, is rightly credited with having forced the subsequent creation of the Environmental Protection Agency and having inspired the ecological movements that took political root during the ensuing decades. There are many today who widely and properly count Rachel Carson a hero.[3]

By today's ecological measure Rachel Carson's scope might perhaps seem narrow, focused as it was on the use of synthetic pesticides, most especially DDT. Yet however limited her focus may have been, Carson unerringly targeted one of the fundamental dimensions of the crisis of the natural life-world. "The chemical weed killers are a bright new toy. They work in a spectacular way;

they give a giddy sense of power over nature to those who wield them, and as for the long-range and less obvious effects—- these are easily brushed off as the baseless imaginings of pessimists."[4] Carson spotlighted the disastrous "disconnect" our presumptive governance of nature helps to promote between immediate applications of instrumental knowledge, usually grounded and tested experimentally, and an entire range of possible outcomes or future events that may follow upon those applications apart from the effects intended. Normally called "side effects" or "unintended consequences," these eventful outcomes may be—in fact, they too often are—not only unplanned and unwanted, but altogether unforeseeable as well. In calling attention to the widespread and disastrous side effects of various synthetic chemicals aimed at eliminating all sorts of pests, Carson voiced what has since become the central tenet of all ecological sensibility, namely the interconnectedness of all things in the natural life-world.

Silent Spring demonstrated how the dynamic interrelationships among living things and between living things and their environments can causally connect through time various phenomena or events that on the surface seem to have little or nothing to do with each other. This insight directly informed the critical ecological consciousness appearing with the success of Carson's work. Barry Commoner enshrined it in his very first law of ecology, which is that everything is connected to everything else.[5] And not too long after Carson's work, but from an altogether different perspective and point of origin, chaos theorists reinforced her insight metaphorically in the so-called "butterfly effect," according to which a butterfly's flapping its wings in Brazil can help to effect a tornado that may later materialize somewhere in Texas.[6] Cause and effect are distanced in time and space in non–linear fashion, with contingency, complexity, and sometimes novelty having opportunity to play various roles in the unfolding of events. Carson disclosed this ecological reality in stories of calamitous side effects following the use of pesticides, most notably DDT. Carson's disclosure of such causality lifts *Silent Spring* above the empirical confines of its data, her empirical case studies simply showing the way of the ecological world. Let us consider the particular case giving her work its poignant title.

Silent Spring took its name from a case study both simple and classic in the story it told.[7] This case was one of several involving the disappearance of songbirds from various locations in the United States in the mid–1950s. Background to the story was itself another story, one going back an entire generation to the inadvertent entry of an arboreal fungus disease to the United States via elm wood imported by the wood-veneer industry. Beyond that story and following upon it, rampant spread of Dutch elm disease had led by the 1950s to the widening use of synthetic chemicals, DDT being the most celebrated "wonder weapon" in the pesticidal armory employed to counterattack the elm bark beetle, the vector of the fungus. Spraying of DDT was widespread and began in moderation on the Michigan State University campus in 1954, where at the time a Ph.D. candidate in ornithology happened to be studying robins under another ornithologist's supervision. This circumstance led to the first awareness of a declining robin population on campus. Carson writes: "Dead and dying birds began to appear on the campus. Few birds were seen in their normal foraging activities or assembling in their usual roosts. Few nests were built; few young appeared. The pattern was repeated with monotonous regularity in succeeding springs."[8] Spring became increasingly silent.

Observation of the birds' death throes and post mortems agreed with the MSU ornithologists' early suspicion that some poison was at work, and, not surprisingly, the obvious culprit was DDT. The question remained, however, as to how DDT was working its lethal effect. The easy answer, and one that might possibly have proved somewhat predictable had it been the case, was wrong: death seemed not to come from any direct contact between the poison and its victims. The investigators then happened upon a piece of good luck when they fortuitously got wind of an event that had taken place in an unrelated research project nearby. That other project involved crustaceans, a number of which had mistakenly been fed earthworms and had died as a result of the mistake. Along this same diagnostic line a snake in a lab cage not far away showed symptoms of DDT poisoning after having been fed the same food. What Carson calls the "key piece in the jigsaw puzzle" then showed up somewhat later with the publication of a study linking together a chain of interactions that, shifting metaphors, filled in the missing chapters of the story.

In abridged form, the story reads as follows: the less than precise application of DDT to the trees had indiscriminately coated leaves and bark everywhere, leaving a layer of poison that had nowhere to go but to the ground upon the coming of autumn. What followed then was the natural cycle in which leaves are turned into soil, a process typically involving earthworms' eating the dead ground cover, which in this case contained DDT. Thus began a process of biological magnification. Each earthworm biologically would concentrate in its body larger and larger amounts of DDT as it continued to eat. Then, when the feeder became the fed-upon, the earthworm would give up its poison to the robin, which would biologically concentrate the poison as it ate earthworm after earthworm until it reached a fatal threshold. As for those birds whose biological variation and meal portions favorably combined to allow their survival, there followed a different conclusion, but one no less devastating to the robin population: high concentrations of DDT in the birds' reproductive organs had compromised the chances of the next generation's hatching. So, in the end, it was truly a case of the brawler's adage — if the right one don't getcha, then the left one will.

Carson submitted her story of the robins' fate in fear of further narratives to be magnified in serial fashion. Along with other bird species, mammals such as raccoons, shrews, and moles eat earthworms, and, of course, shrews and moles themselves occupy a choice place on the menus of predatory birds such as owls and hawks. And so life goes, or fails to go, up and down the food chain. So, too, DDT was perhaps only the most infamous of the numerous synthetic chemicals Carson addressed more generally in her concern to expose the widely diffuse pesticidal poisoning of the environment and the carcinogenic threat it posed to human well-being. Other stories were in the making that Carson herself would not have been surprised to read had she lived, and we shall encounter a couple of them in what lies ahead. Before moving on, however, we need to highlight what the simple case of the silent spring fundamentally reveals about the ecological reality in which we are enmeshed and into which we instrumentally insert ourselves too often as blunderers.

DDT's effects on hosts of different pests were well established experimentally before mid–century. Resistance to pesticides already was underway in the 1950s, and Carson warned of the continued large-scale use of pesticides leading to "super bugs." But here we are not concerned with arthropods' resistance to pesticides, a problem that has grown to immense proportions over the past half century.[9] In the present context we grant DDT the same power to eliminate all sorts of arthropods that experimental testing had demonstrated prior to patenting and marketing. The critical point here is that the myriad possible side effects of applying DDT were not specifiable upon the observational conditions given with the testing. Most basically, the difficulty with ecological side effects is that one cannot know beforehand precisely all of the interactional variables to be tested; hence, possible problems are not altogether specifiable. Nor does moving testing from the laboratory to the field necessarily identify all possible or relevant side effects, for ecological contingencies can and do differ from field to field. Spring did not become silent everywhere that DDT was used in large measure. Or, as Platt and Griffiths have written more generally: "Plant, animals, and other entities exist in uncontrolled environments, and we can only partially control, at best, the complex of interacting factors which make up the natural environment.... For experimental procedures in natural environments, the laboratory concept of reproducibility has little validity, inasmuch as nature never produces exactly the same conditions...."[10] The problematic side effects of birds' dying or becoming reproductively disabled became identifiable only once these effects had already appeared. At that point, of course, those ecological casualties constituted a *given*, and causality became a question after the fact. Given the side effects, facing the victimized birds, scientific investigators like good detectives then had to piece together bits of evidence to reconstruct the crime. Such reconstruction necessarily was historical, and the non–linear, interactional causality fashioned by the investigators took the form of the story Rachel Carson retold in *Silent Spring*.

The case of silent spring clearly and simply illustrates the interactional difficulties contingency and complexity often pose for prediction. Contingency and complexity pose the very same problems for explanation as well. Explanation of the robins' deaths did not take shape overnight, but the case of silent spring was fairly straightforward, at least with regard to the culprit. Identifying DDT as the criminal did not take long at all. The decided advantage of early detection is, however, often not the case. While the renowned French method of reconstructing the crime may predictably pay dividends in detective fiction, in the real world of ecological affairs the realm of appearance does not follow any script, and identifying the culprit not uncommonly presents one of the most challenging difficulties. Do we have a suspect at the start? And if we do, how do we establish guilt or innocence when the guilty party, whether it be the suspect or not, may be hiding evidence within the complexity or behind some contingency? Questions of this sort stand out strikingly in the contemporary case of colony collapse disorder, or CCD, which regards the disappearance of the honeybees.

Colony collapse disorder designates a phenomenon that has been recorded in various countries for over a decade now. *Collapse* refers to the honeybee communities' sudden disappearance, as some insidious disease and/or radical behavioral abnormality has caused the bees to abandon their hives. They simply do not return home. Why this phenomenon is occurring in various parts of the world, including the United States and parts of Europe, is a most serious and urgent question because CCD threatens our agriculture in fundamental ways. Average loss of hives reported by beekeepers surveyed in the United States amounted to about 32 percent from 2006 to 2007; about 36 percent from 2007 to 2008; and about 29 percent from September 2008 to April 2009.[11] Numerous nuts, berries, fruits, and vegetables depend on honeybee pollination, and about one-third of the agricultural crops in the United States currently could be endangered. There is no guaranteed protection against this risk because no solution to the problem is clearly in sight, at least not to the satisfaction or agreement of the investigators. Indeed, even identifying precisely the nature of CCD has been problematic. One of the initial problems of investigation sprang from the simple fact that, since the bees fail to return to the hive, the victims' corpses were not plentifully available for post mortems. Another problem is knowing whether or not CCD is novel, and determining this issue may be impossible. Most fundamentally, of course, all problems of investigation ultimately relate directly or indirectly to identifying the nature of CCD and, with this, to finding out what is causing it.

Beekeepers in France identified widespread CCD in the mid–1990s, roughly a full decade before it reached any noticeable or alarming magnitude in the United States. It is not clear, however, whether or not colony collapse disorder had ever been seen before. For over a century beekeepers periodically have reported cases of hives being abandoned suddenly, using such terminology as "disappearing disease," which is on one record from Utah dated 1903.[12] However, as the United States Department of Agriculture's Research Service is quick to point out respecting any and all earlier records, "there is no way to know for sure if the problems were caused by the same agents as today's CCD."[13] Some beekeepers and researchers, nevertheless, believe CCD was occurring forty or fifty years ago and that it did not appear as such simply because it lacked epidemic proportions. Other researchers, probably the majority, believe to the contrary that it is attributable to new environmental elements acting singly, combinatorially, or synergistically. Neither beekeepers nor scientific investigators are agreed among themselves on this issue, and other issues as well divide those who have been tackling the question of CCD's causality. Among those various issues, the one standing out for our own purposes concerns the effects of certain neonicotinoid pesticides, most especially imidacloprid, or IMD. Is CCD another case of catastrophic pesticidal poisoning?

Environmental contamination is unfathomably deep and pervasive today, and since Carson's time pesticides, to name but one class of contaminants, have proliferated without interruption in the spiraling conflict between our chemical weaponry and arthopodal resistance. Given the criminal record of pesticides in general, suspicion of IMD arose at the very start of serious investigation of the honeybee disorder.[14] French beekeepers were quick to convict the pesticide. The agrochemical industry, on the other hand, has suffered no uncertainty whatsoever about the innocence of their product. To hear them tell it, their testing of IMD has unfailingly continued to exonerate this chlorinated nicotine pesticide that, in its neuro-toxic effects, disables the very capabilities bees need to return home to their hives. Being employed around the globe to combat various pests in orchards, crop fields, lawns, and golf courses, IMD is not only profitable but widespread as well. Over and against the agrochemical industry's powerful defense of their product, the French government nevertheless banned the use of IMD, along with another neonicotinoid, fipronil, in 2005. Behind that decision lay a significant experimental investigation of IMD's effects on bees exposed to sub–lethal doses, carried out the year before by a team of French scientists.[15] What proved conclusive enough for a condemnatory verdict in France has, however, yet to persuade those who count elsewhere. Outside of France various continuing lines of investigation have been pursued alongside those looking into IMD. Some investigators are targeting different pathogens and parasites while others are looking into various environmental stress factors, especially those relating to the handling and transportation of hives. To date, in early 2010, neither a majority of researchers nor a majority of beekeepers have reached any absolute consensus. While individuals may have their own strong beliefs or hunches, the jury as a whole remains out.

Failure to reach a verdict on IMD follows from more than just the power of the agrochemical industry, or from at least the possibility that the defense could have bought off a few of the jurors through private contracting or institutional funding affecting their research findings and careers. Uncertainty is a genuine condition for the large majority of scientific investigators, even in the face of some seemingly strong prosecutorial evidence against IMD. French beekeepers offered Exhibit A when they reported a tremendous resurgence of honeybee populations in 2006 and 2007, directly following the ban on IMD and fipronil. This restoration of normal health would appear to carry some serious evidentiary weight. The problem, however, is that this return to the status quo ante was not uniform across France. Even in the absence of IMD there proved to be little or no buzzing in those areas of France that happened to be suffering drought. One might submit that the inhospitable conditions of drought worked against the bees' rebounding in such places, which is just what French beekeepers did argue.[16] But this argument runs the risk of actually weakening the case against IMD. Arguing that there would have been complete recovery from CCD in France were it not for regional differences may attenuate the case against IMD, or at least resurrect some legitimate uncertainty, by turning what seemed at first to be *prima facie* evidence into contingently circumstantial material.

Circumstantial contingencies contribute to uncertainty about IMD in other situations as well. The prosecutors' case against the pesticide appears to face serious difficulty in Australia and New Zealand, which both permit the use of IMD but have yet to witness any sign of colony collapse disorder. Once again contingencies are called upon by those who believe IMD guilty, or at least complicit, in CCD. In this particular case the prosecution must explain the absence of any hive disorder despite the presence of the suspect. In these lands of Oceania, integrated pest management, tight biosecurity laws, and other factors have resulted in fewer pests and much reduced, more stringent applications of the chemical. These factors may be viewed together as an explanation for why hive communities have escaped CCD in the presence of the suspect pesticide, but such "evidence" tends to assume in the circumstances what remains to be established in fact. Investigative science writer Michael Schacker in any case concludes from all this that "[i]f IMD truly is the prime suspect, Australia and New Zealand would serve as models to immediately follow."[17] This point carries weight in the interest of prudence, and there is everything to be said for circumspect sensibility. That said, the jury, nevertheless, needs remain uncertain in the face of such circumstantial evidence. Ideally, and quite understandably, they would prefer a smoking gun.

Schacker, whose book is tellingly titled *A Spring Without Bees*, is one of those who suspects that IMD may well be the guilty party, or at least directly complicit, and he points to a study done by Chris Mullin of Penn State, which shows traces of IMD and other toxins in virtually all samples of beeswax and pollen tested from abandoned hives.[18] Schacker is hopeful that this study along with IMD experiments carried out by Marc Colin will persuade other investigators to pursue similar lines of investigation, and the investigators he especially has in mind are those who have questioned the consistency of any research findings regarding the role of pesticides in CCD. One such researcher has been entomologist Jeff Pettis, a leading figure in the Agricultural Research Service of the USDA and one of the most widely quoted investigators of colony collapse disorder. Schacker seems to imply that Mullin's study and Colin's experimental work on IMD might be pointing toward the smoking gun that scientists like Pettis need in order to arrive at some conviction. But Pettis opposes the very idea of a smoking gun: "What I believe is that CCD is likely a combination of factors, as opposed to a single cause...."[19] Nor is he alone in such thinking. Speaking for a great many scientists in the field, Chip Taylor, a bee specialist at the University of Kansas, says, "We believe that what we're going to find is that it's not a single factor ... but that there are multiple factors involved.'"[20] Even Mullin himself looks for a group of operators: a toxic gang could be working together, perhaps under the leadership of IMD but acting synergistically.[21]

We have not been spotlighting IMD in the interest of joining the prosecution team. As a single, clearly identifiable factor, the pesticide simply has provided a focal point from which to survey a few of the difficulties that contingent interactional dynamics can pose for ecological investigators. In point of fact, IMD was not directly implicated in an important genomic study undertaken by the CCD Steering Committee set up in 2007 by the Agricultural Research Service along with another USDA educational service. Presenting findings from its work in progress in June 2009, the Steering Committee's genomic comparison of CCD victims and healthy bees discovered various viral culprits that were attacking the bees' ribosomes, and suggested that these viruses likely have success where bees' resistance to them has been weakened by environmental stress.[22] This study as a whole may be very promising, but in and of itself it has neither exonerated IMD nor led to the conviction of any particular group of suspects. Unlike the situation we had with silent spring, the uncertainty we find in the case of CCD is current. It remains to be seen whether or not new or ongoing investigations will in the near future erode this uncertainty and lead to a clear verdict of some sort. Time will tell in any case. And in the meantime, much of our agriculture will remain at risk.

Our discussion of silent spring featured the problematic nature of predicting the side effects of our applied knowledge, and the discussion of CCD has highlighted different environmental contingencies that can bedevil the investigation of causality and, with this, the explanation of some already given effect. Environmental interactions have been the critical factor in both cases. Interactions take time to unfold, of course, yet in neither of these two cases has temporality itself been the major problem. We want now to turn to two different cases that magnify the interactional problem of prediction through the temporal lens of ecological affairs. Each case from a particular angle features the element of time in the reality of delayed effect. In the first case, which we will examine in the section to follow, all of the proper scientific testing had been performed on a chemical that would be employed widely in people's homes and in industries in various technologies. No potential hazards of any kind had been detected, and things appeared to be just fine — not only at the time of the testing, but, moreover, throughout the half-century of technological employment that was to follow. This case is the now infamous story of chlorofluorocarbons, or CFCs.

Interactionally Delayed Effect

Chemist Thomas Midgley Jr. in 1928 synthesized the first chlorofluorocarbon, initiating a long line of laboratory products quickly developed by industrial chemists in the early 1930s. Modern refrigeration was in its infancy, and CFCs were coolants having immediate application. CFCs also were used as propellants and found immediate employment in foam and insulation, and in the cleaning of industrial equipment. Here was a classic instance of new instrumental knowledge having wide application in a number of industries at the forefront of twentieth-century progress. All such applications of course presupposed that CFCs are to all intents and purposes harmless, which is just what all of the testing had shown prior to any commercial use. These synthetic chemicals proved to be nontoxic, nonflammable, noncorrosive, and nonreactive in the air we breathe.[23] Quite simply, they were benign. Midgley himself showcased the safety of his carbon compound with dramatic flair at the meeting of the American Chemical Society in 1930, where he inhaled vapors of his liquid chemical and then extinguished the flame of a burning candle by exhalation.[24] The ultimate proof, or confirmation, of the harmless nature of CFC's came with their various domestic and industrial employments over the next forty years. There was nothing in the world wrong with this synthetic chemical, and certainly no one had any reason to suspect any danger anywhere. But that assessment changed for truly alarming reasons, beginning with a discovery made at the start of the 1970s.

Most people likely connect James Lovelock's name with his Gaia hypothesis, which concerns planetary understanding, but Lovelock's most consequential work unquestionably pertains to CFCs. Indeed, it was his discovery of their presence in the atmosphere that gave others reason to search out their effects. Lovelock had invented the electron-capture gas chromatograph, an instrument allowing him to determine aerosols (extremely small airborne particles) in the atmosphere, and what he found were CFC molecules everywhere: first in the air he measured throughout the region of his home in the British Isles, and subsequently in all of the atmospheric samples he took aboard an expeditionary ship traveling between Great Britain and Antarctica.[25] Lovelock did not himself follow up his discovery, however. The scientific investigation of CFCs in the atmosphere fell almost immediately to the chemist F. Sherwood Rowland, who, along with Mario Molina and Paul Crutzen, would later be awarded the Nobel Prize in chemistry for work in atmospheric chemistry, more particularly for his investigation of chlorofluorocarbons. As Rowland reflected in his Nobel lecture in 1995: "The appearance in the atmosphere of a new, man-made molecule provided a scientific chemical challenge: Was enough known about the physicochemical behavior under atmospheric conditions of molecules such as CCI3F to allow prediction of its fate, once released into the environment?"[26] Investigating that question led Rowland and his two co-recipients of the Nobel Prize to launch a global alert. Alert, in fact, may be far too weak a word to convey what these researchers actually felt about their findings in the midst of their investigation. Elizabeth Kolbert quotes Rowland's telling his wife at one point, "'The work is going well, but it looks like it might be the end of the world.'"[27]

Rowland and Molina discovered that CFCs hang around in the troposphere without any ill consequences for very long periods of time. Their stability and nonreactive nature give them a lifetime of upwards of 100 years, which on the face of it seemed nonproblematic. After all, we had been breathing them and using them industrially and domestically for decades. But the problem Rowland spotlighted lies not in our world, at least not in the troposphere. The problem lies in the upper atmosphere, the stratosphere. Being heavy, CFC molecules are quite content to stay near to the earth; these molecules do not migrate upward on their own. Subject to various atmospheric pulls and pushes, however, many of these homebodies are forced to leave. Those leaving do so reluctantly, and it takes anywhere between one and two decades for them to reach the stratosphere. Then, once these CFC molecules are in those inhospitable environs, they lose their stability. These synthetic compounds break down and produce bad actors whose molecular behavior attacks the ozone, that gaseous form of oxygen which makes up a thick stratospheric layer that, like a shield, protects the rest of us living back home from the sun's ultraviolet radiation. Much life on this planet, including our own, could face disastrous survival problems without this shield, which prevents 99 percent of total solar UV radiation from reaching Earth. It requires little imagination to appreciate the extreme concern Rowland spoke to his wife.

Looking at CFCs more directly, they are organic compounds in which hydrogen atoms are removed and replaced by the halogens chlorine and fluorine.[28] When these compounds reach the stratosphere, they are subject to the very same ultraviolet radiation against which the ozone protects us. CFCs do not readily withstand such high energy radiation, and sooner or later, depending upon the altitude or stratospheric layer, the UV-induced dissociation of CFCs' molecular structure frees up the halogens. Chlorine and fluorine radicals emerge and steal oxygen atoms from ozone molecules, fostering new compounds while destroying the ozone. Chlorine catalyzes the chemical destruction of ozone molecules, meaning that it remains intact, moving from one ozone molecule to another like some unstoppable destroyer. Fortunately, other chemical reactions take place that take out the destroyer, but usually not until these attackers have been able to wreak some real damage to the ozone. This damage is what we generally read about when scientists write of ozone depletion. More commonly, perhaps, we hear about the "ozone hole," which is a special case of ozone depletion appearing at, or over, the poles, most remarkably in Antarctica, where regionally exceptional temperatures and various seasonal phenomena cyclically thin out the ozone to the point where a hole can be visualized by computer imaging each spring.[29] The global depletion itself is what counts as a threat, but the discovery of an ozone hole over the South Pole in 1985 has proven important to environmental concern and, most crucially, to the general development of governments' sensibility around the globe.[30]

Rowland and Molina's research led them in 1974 to call publicly for a ban on spray cans, as these contained propellant CFCs. Thus began a struggle between those seeking action first to limit and then to ban chlorofluorocarbons and those who saw no good reason to alter the status quo. Donald Hodel, at the time President Reagan's secretary of the interior and one of the more environmentally astute cabinet members in Reagan's altogether environmentally astute administration, proposed three simple solutions to any and all problems posed by ozone depletion: buy a hat, wear sunglasses, and do not stand in the sun.[31] While Hodel was being glibly dismissive from a position of some public prominence, behind the scenes the chemical industry quietly but all too predictably waged its own campaign to dampen concern by denying that CFCs had anything to do with stratospheric events, whatever those events might or might not be. Leading the industry's cause in the name of the chlorofluorocarbons' innocence were companies like DuPont and Allied Signal, an unsurprising leadership given the long record these companies had of championing both the common good and the disinterested pursuit of truth in public affairs. At one point the chair of Dupont's board was quoted as calling the argument for ozone depletion "science fiction" and "rubbish," which demeaned both science and science fiction in one foolish, self-serving judgment.[32] The chemical industry's efforts persisted into the late 1980s, even up through the time that a smaller ozone hole appeared over the North Pole. But those efforts proved to no avail against the mounting evidence and growing concern. Ozone depletion was altogether too real and posed threats of the very gravest sort.

Observational evidence supported the predictions coming with both Rowland and Molina's work and the research following theirs. Chemistry transport models auguring dire results from continuing CFC use proved correct in view of empirical findings, and, of course, the discovery of the ozone hole dramatized in the public's mind the issue of ozone depletion. Depletion could hardly be stopped all at once because many of the CFCs already in use would not remain earthbound, and so the overall longevity of CFCs in the stratosphere was daunting, to say the least. The various chemical chain reactions that cyclically eliminate and replace ozone are negatively out of balance, and scientists have calculated that this negative imbalance will grow for some time but that the ozone layer could be restored to its 1980 level in 50 years, and, by the end of this century, its comeback might reach pre–1950 levels.[33] All such figures assume political action to rein in the production and use of CFCs, which, of course, was the very action the chemical industry feared and opposed for so long. Opponents of political action might well have won the day had the consequences of inaction not been potentially disastrous.

Ultraviolet radiation comes in three different sizes, or wavelengths, and the ozone protects us from the worst kind, UV-B. Sunburn results from the UV-B radiation reaching Earth, as does skin cancer of various kinds, and the likelihood of skin cancer correlates directly with the incidence and severity of sunburns. Indeed, experimentation has disclosed the chemical mechanism by which UV-B causes skin cancer, and 90 percent of all skin carcinomas are attributable to UV-B exposure.[34] All racial pigmentations allow sunburn, but Caucasians are far more susceptible to sunburn than are people of color, and they suffer a considerably higher incidence of skin cancer. It is exposure that counts most in any case, of course, and estimates at this point are that a 1 percent decrease in the ozone shield likely will effect a 2 percent increase in UV-B radiation, which, in turn, likely will effect increases of 4 percent and 6 percent in the two less threatening kinds of skin cancer.[35] In any event, human harm, existing or potential, goes beyond cancer, as serious eye damage may also result from increasing exposure to ultraviolet light. But there is another threat, which in theory at least is yet the gravest of all. This potential threat actually targets us indirectly, as it comes to us and to all other advanced, more complex life species by way of the lower levels of the organic kingdom.

Reagan's secretary of the interior may have been right with respect to our own immediate contact with UV-B—if the sun begins to pose a serious threat to our well being, we need to stay out of it as best we can—but advice of this kind has nothing whatsoever to offer other life species lacking the intelligence and choice-making of Hodel's own. They may be considerably less adaptive in their capacity to escape the sun, as would seem to be the case with phytoplankton. We never were told anything about phytoplankton in those old, traditional discussions of the Great Chain of Being, yet these microorganisms are involved with generally half of the phosynthesis occurring on earth and fix much of the world's CO_2 in organic forms.[36] They also are the foundational producers at the very bottom of the oceans' food chain. These tiny little organisms are to the oceans what green plants are to the land: they fundamentally sustain all life in it. Phytoplankton are the food on which zooplankton feed, giving these primary consumers their sustenance; zooplankton then become food for secondary consumers, giving small fishes their essential diet; small fishes, in their turn, become prey to the larger ichthyophagous consumers, and so on. So it goes all the way up the ocean's food chain to the shores beyond, where so many species, including our own, depend in no small part on the oceans' bounty, rapidly depleting though that may currently be.

Although we need to think in terms of nonlinear interactional webs when addressing ecological relationships, as well we should, what we are talking about here actually resembles a house of cards. Pull out a bottom card, and everything collapses. Marine biologists are not now predicting any ecological calamity beginning with plankton, but there is concern for these microorganisms that drift within two meters of the oceans' surface and also float at the top, where exposure to UV is greatest. Although there is nothing conclusive in any of the research carried out thus far, most phytoplankton have shown some considerable sensitivity to increased levels of UV-B.[37] So, too, nitrogen-fixing soil microbes on which land life depends so much, have shown vulnerability to UV radiation. But enough said on these matters. Scientists know too little to venture any reasonable guesses about any serious damage that might follow from continuing ozone depletion. Fortunately, the effects of UV-B radiation on our skin were known well enough, and the magnitude of the general ecological dangers were sufficiently large enough, to bring together in 1987 the governments of all the major CFC-producing countries, except China, to deliberate and act on the ozone crisis.[38] The world, or a good part of it, was ready to take political action.

These governments met in Montreal and worked out an agreement, the Montreal Protocol, which was a political milestone in showing what government leaders can achieve internationally when they sense a clear and present danger commonly threatening their self-interests. Countries' self-interests require some such common ground for global action to occur where global sensibility is lacking—a political truism all too apt for our continuing era of nation states. Critical to action on CFCs was the widespread conviction that ozone depletion constitutes a clear and present danger to everyone. Probably just as important as this conviction, however, was the fact that replacement technologies were either coming on line or likely to be available in the near term. Air conditioning would not disappear, and the chemical industries had new profit fields to plow; major considerations no doubt permitting many government leaders the clear vision necessary for action. The Montreal Protocol set an upper limit on CFC production at the 1986 level and mandated a 50 percent reduction in the production of CFCs by the year 1999.[39] Two subsequent agreements quickly followed this one, as new ozone assessments further darkened the stratospheric picture. The London meeting in 1990 substituted 100 percent reduction for 50 percent, and then, two years later in Copenhagen, world leaders moved up the year for zero production from 1999 to 1996. These agreements could not immediately eliminate ozone depletion, but what they did accomplish is perhaps immeasurable. Although we have no control case to hold constant so as to observe how things might have played out had the production and use of CFCs continued unabated, at the very extreme end of speculative possibilities we might imagine with the virtual destruction of the ozone a mass extinction to rival what occurred at the so-called K/T boundary of geologic ages when, some 65 million years ago, upwards of 70 percent of all species disappeared in the wake of a massive asteroid's unasked-for arrival.

Global agreements on CFCs turned on major political figures' not buying the sales pitch of those labeling ozone concerns rubbish or science fiction. The political leaders' actions rested, in turn, upon the efforts of scientists who followed the thrust of their findings into the public arena, where they spoke and acted with a responsibility and urgency proper to both their professional knowledge and their ethical sensibilities. At bottom, however, what made everything possible in the first place was a wonderfully fruitful wedding of happenstance with circumstance. Lovelock had just happened to invent a piece of equipment that could and did capture CFC aerosols in the lower troposphere. This invention made possible everything else that followed, though not directly. The electron-capture gas chromatograph did not lead Lovelock to follow up his own finding. That follow-up fell to Rowland, who made use of the circumstances provided him by Lovelock's work. Just as important to the possibility of Rowland's own work was the U.S. Atomic Energy Commission's decision to allow Rowland to shift some of their funding of his other work to his proposed investigation of the atmospheric effects of chlorofluorocarbons.[40] Other matters also came into play, including Paul Crutzen's work in the atmospheric transport of chemicals. The critical point here, however, is that Rowland seized an opportunity opened up adventitiously. At the very foundation of what ultimately may have had immeasurable significance for humankind and other species, there lay an enormous amount of good fortune.

Crutzen also has spoken of the good fortune lying at the heart of the CFC story, but his point has a different slant. Reflecting upon the ozone situation in that same widely influential piece he wrote for *Nature*, Crutzen offered this deeply disturbing thought: "Things could have become much worse.... If it had turned out that chlorine behaved chemically like bromine, the ozone hole would by then have been a global, year-round phenomenon, not just an event of the Antarctic spring. More by luck than by wisdom, this catastrophic situation did not develop."[41] What strikes us most forcefully here are two matters: the great luck obtaining in the fact that chlorine could not wipe out the ozone as bromine would have, and the truly alarming fact that, could it have, there would have been little or no reasonable possibility of our knowing its effect until that enormity was already upon us. In this last regard, time worked in the chemical's favor, so to speak, for whatever its stratospheric effects might prove to be, they were delayed by the nature of the chemical, whose behavior in the meantime was nothing but benign.

Delayed effect followed upon interactions lasting decades. Decades-long interactions were possible in the first place, of course, because these synthetic compounds hang around for a very long time. Nothing dangerous suggested itself during that long period; indeed, everything we ourselves could observe in widespread industrial and domestic venues only confirmed the benign nature of the chemical that laboratory experimentation had verily determined at the outset. What arises here in this case of delayed effect, therefore, is irony in the realm of appearance. While nothing untoward appeared biologically or ecologically in either the experimentally controlled or the everyday realms of appearance, ecological interactions themselves were readying a causality with the patience that only nature can exercise. And then, after decades, when scientists actually were able to see that causality at work, it had already produced the makings of a story that was well underway; a story studded with unwanted events and having a potentially catastrophic conclusion. Good luck favored us nonetheless. Good luck did not, however, move the scientists to demand that we politically insert ourselves into the story in order to radically alter its script: ethical sense and prudence did that.

Laplaceans among us should find absolutely no comfort in the stories we have told, or retold, thus far. If, however, we allow super vision to be assumed in the most highly idealized fashion and extended in the most unreasonable could've, might've vein, then perhaps some Laplaceans will still insist on the possibility of accurate forecasting under conditions of completely thorough and thoroughly complete knowledge. Such thinking is, after all, Laplacean. But nature itself plays no small part in determining when, where, and under what conditions it will allow scientists in the real world to see its causality at work. Already we have seen in the biological kingdom various instances of this reigning fact. Now we want to look at a case that combines the ecology within our bodies with the ecology outside our bodies in a particular dynamic that ought to silence any and all residual Laplacean whispers. Like other stories we have retold, this one speaks in representative terms: it represents a whole host of similar cases that exist all around us and, quite likely, within us as well. And like the story of CFCs and the depletion of ozone, this case points to possible eventualities deeply disturbing and long delayed in their appearance.

This particular story spotlights in our own organic life-time causal dimensions of organic change and development that plainly put certain, predictive knowledge out of our reach, pure and simple. We retell this story primarily from the book *Our Stolen Future: Are We Threatening Our Fertility, Intelligence, and Survival?— A Scientific Detective Story*. This book is not uncommonly judged the finest ecological work since Rachel Carson's *Silent Spring*. It is a book about lifetime: ours, our children's, and our children's children as well.

Organically Delayed Effect

Carson clearly was aware of the harmful effects DDT and other chemicals might have upon various species' reproductive capabilities, and had she lived, she would have been anything but surprised by the publication in 1996 of the book *Our Stolen Future: Are We Threatening Our Fertility, Intelligence, and Survival?— A Scientific Detective Story.* Written by two scientists and a science writer, this work exposed many of the reproductive threats chemically at large in the environment.[42] Called in the most telling fashion a scientific detective story, *Our Stolen Future* accumulated and synthesized evidence from a large number of case studies and scientific experiments covering the last five decades of the twentieth century. Al Gore, writing a "Forward" to the work, said that in many respects it was a sequel to Carson's *Silent Spring*, and he generally was not wrong in this assessment.[43] Gore's judgment might, nevertheless, have struck the book's authors, especially lead author Theo Colborn, with more than slight irony since they believed *Silent Spring* in its own way to have had a deleterious effect on ecological study. Without denying the tremendous significance of Carson's work, Colborn saw in that work's actual power and success a skewing of ecological concern and attention away from the very material that only much belatedly, in 1996, would find its way to public light in *Our Stolen Future.*[44]

Carson's focusing so forcefully on the carcinogenic effects of synthetic pesticides tended to fix attention on cancer-causing chemicals and their disease-threatening potential for human sickness. Largely overlooked in the swirl of controversy following her work was the endocrinological ecology within our bodies and the hazardous potential myriad synthetic chemicals pose to the hormonal constituents of our health—harm threatened not in the form of some debilitating or fatal disease such as cancer, but, rather, in the manner of hormonal disruption and developmental or hereditary abnormality that may show up at different times of life. Carson duly remarked on reproductive problems in her work, but she never pursued the hormonal pathways leading to such problems. Moreover, her own critical interest narrowed as she unfolded her critique, and *Silent Spring* ultimately became an indictment of technologies that might be linked to cancer. That work certainly defined humankind's fundamental predicament as an uncontrolled experiment on human well-being undertaken blindly by human beings themselves. What *Silent Spring* failed to convey fully or properly was the vast reach of that experiment and the enormity of the stakes on the table of our experimental gamble. Understanding of that larger picture, or at least a significant portion of its chemical constitution, took decades to find more complete articulation. That achievement largely was Colborn's.

Not all crimes are the same or equally evident. As Colborn notes, "It was hard to miss the masses of dead birds that littered suburban backyards after aerial spraying in the1950's.... Abnormal parental behavior or poor survival of young, on the other hand, are less immediately apparent but perhaps no less important in the long run to a species' survival."[45] Birds' DDT-ridden corpses and animals' abnormal behavior or abnormal development evidence different kinds of crime, and Carson's work had helped to identify and circulate the *modus operandi* of the chemical criminals causing disease and death. In order to undertake her own successful criminal detection, Colborn's understanding would have to break away from the disease and death m.o. preoccupying the criminal investigations of ecological problems everywhere at the time. Her own breakthrough would occur only when she was able to identify the m.o. particular to the chemicals lying behind a number of seemingly unrelated crime stories reporting biological abnormalities in globally disparate geographic regions. Connecting those stories by means of their common m.o. ultimately would identify a new class of chemical criminals.

The opening chapter of *Our Stolen Future*, entitled "Omens," contains what are essentially news flashes, each followed by a news brief.[46] As these news flashes appear: "1952: Gulf Coast, Florida.... The late 1950's: England.... The mid–1960's: Lake Michigan.... 1970: Lake Ontario.... The early 1970's: The Channel Islands, Southern California.... The 1980's: Lake Apopka, Florida.... 1988: Northern Europe.... The early 1990's: The Mediterranean Sea.... 1992: Copenhagen, Denmark." And the briefs are no less wide ranging in their content: in one locale, bald eagles declining in numbers as adult eagles fail to follow courtship and mating rituals; in another locale, otters disappearing from fields and streams; mink ranchers in one area reporting that female minks were not reproducing; a wildlife biologist reporting in another area that the unhatched chicks of gulls exhibit gross deformities; another species of gull twenty-five hundred miles away evidencing same-sex nest sharing; an alligator population's decline perhaps linked to the majority of its males having extremely small penises; huge die-offs of seals and dolphins under seemingly suspicious environmental conditions; and last but certainly not least, a report of human sperm count having fallen 50 percent and testicular cancer having tripled among Danish males between the start of the World War II and 1990. Except for this last piece of news, nothing taken by itself necessarily captures our attention in this catalogue of ecological crime stories. But once these events and others are seen to form a pattern, they become alarming indeed. That patterning is the scientific detective story Theo Colborn tells.

The background to the story is that, having been reported locally, the crime stories already were in the books. The problem Colborn faced was linkage: most immediately, at the local level, the connections to be made between specific criminals, most notably PCBs, DDT, and dioxin, and the various crimes reported; and, more generally, the endocrinological linkage of these disparate cases from varying environmental venues. Colborn's own professional journey along the connecting route started with her looking for carcinogens in the Great Lakes area as part of a research project in 1987.[47] Having determined that cancer in the area was not a particular problem, she then began searching for a different way to address the threats of the biologically magnified poisons she found in the upper reaches of the food chain. Studies carried out by two researchers working for the Canadian Wildlife Service in the Great Lakes area were of great assistance, as was the connection they provided her to the work of a toxicologist in California whose investigations of the sexual effects of DDT and other synthetic chemicals on birds had led to the discovery of hormonal disruption that caused various behavioral and reproductive abnormalities. Colborn herself had to study endocrinology at that point, and her learning of the ability of some synthetic chemicals to mimic estrogen became the key in her discerning a pattern of hormonal disruption underlying those particular news briefs noted above, along with hundreds of other stories of abnormalities. This key concerns chemical messages within the body's internal ecology.

Those chemicals called hormone-disrupting imitate the body's own hormones, interfering with estrogen, testosterone, thyroid metabolism, and the immune system, as well as a host of other endocrinological matters. For example, by mimicking estrogen foreign chemicals like DDT and dioxin can fool the body by appearing natural, thereby gaining entry to delicate biological processes that they then can disrupt. Viewed environmentally or from the standpoint of adult health, the level of toxicity for a given hormone-disrupting chemical may be far from worrisome. But the perpetrators in this class of criminal chemicals do not intend to victimize adults. Rather, they target the lives taking developmental shape in adult women's wombs. Critically at stake are the developing fetuses; theirs is the delicate endocrinology lying vulnerable to the chemical messages foreign messengers can scramble by way of their alien language. This metaphor belongs to the authors of *Our Stolen Future*, who speak of the criminal chemicals as follows:

> Their target is hormones, the chemical messengers that move about constantly within the body's communications network. Hormonally active synthetic chemicals are thugs on the biological information highway that sabotage vital communication. They mug the messengers or impersonate them. They jam signals. They scramble messages. They sow disinformation. They wreak all manner of havoc. Because hormone messages orchestrate many critical aspects of development, from sexual differentiation to brain organization, hormone-disrupting chemicals pose a particular hazard before birth and early in life....The process that unfolds in the womb and creates a normal, healthy baby depends on getting the right hormone message to the fetus at the right time. The key concept in thinking about this kind of toxic assault is chemical messages. Not poisons, not carcinogens, but chemical messages.[48]

This conceptual shift speaks directly to the critical difference between the m.o. by which hormone-disrupting chemicals are to be identified and that by which standard poisons and carcinogens operate. For example, synthetic chemicals that disrupt hormonal development do not comply with the typical toxicological model that Carson had followed, according to which increasing dosages intensify or enlarge effects. Working differently, synthetic hormone-disrupters are therefore not about to be captured by the experimental devices used to detect the other class of criminals; and, historically, as long as laboratory testing followed what the authors call the "cancer model," criminal detection was limited and in many respects bound to fail at the very start of experimental testing. Colborn writes: "Since hormone-disrupting chemicals do not play by the same rules as classical poisons and carcinogens, efforts to apply conventional toxicological and epidemiological approaches to this problem have typically led to more confusion than enlightenment."[49] In part, then, controlling variables critically relevant to predicting significant effects failed to happen — not for want of experimentation, but from ignorance of both what was to be tested for and what the proper experimental approach should have been. But the ignorance we suffer respecting the effects such chemicals may have on our bodies, or those of our offspring, can go even further than the discussion above suggests.

Economic and political interests all too often limit experimental questions pertaining to the testing of new technologies, but the point here is that plain ignorance itself can limit the nature and reach of such experimentation. Broadly considered, hormone-disrupting chemicals illustrate the critical significance of our needing to know the right kinds of questions to ask beforehand, but these chemicals are most especially instructive in so far as they illuminate our ignorance as an *inescapable* condition. The causal course of hormone disruption takes place behind our backs, as it were, operating with a stealth and invisibility making it absolutely undetectable through long stretches of lifetime. One classical case of a chemical's long-delayed action may spotlight the point best. This case involved diethylstilbestrol, or DES, and came on the heels of the more sensationalized thalidomide tragedy of the early 1960s, which had involved a DNA-inhibiting drug rather than a hormone disruptor, but which did display absolutely horrible birth defects in thousands of cases reported from forty-seven countries.[50] Pregnant women taking the prescription drug thalidomide to ease the stress of pregnancy had wreaked all kinds of damage upon their unborn children, and when the sensationalist tabloids presented their visual display of "thalidomide babies" having hands emerging from shoulders and trunks without any arms or legs, the world got its first real glimpse of what the effects of "chemical wonders" can look like when things go badly awry. From our perspective, however, DES is the much more informative case. As Colborn tells us, "If thalidomide exploded the myth of the inviolable womb forever, the DES experience toppled the notion that birth defects have to be immediate and visible to be important."[51]

DES was a synthetic estrogen developed in the late 1930s and immediately prescribed to help prevent miscarriages and lessen the likelihood of premature births. Twenty years later medical science was recommending it for all pregnancies. As if following the belief that one cannot get too much of a good thing, medical professionals had come to view DES as a biological enhancement supplement for pregnancies in general. Then, in the early 1970s, after thirty years of prescription and increasing use, DES directly was implicated in the etiology of a rare form of cervical cancer that had appeared in a cluster of cases in the late 1960s. The mothers of the large majority of adult women suffering from this rare cancer had all used DES while pregnant. Studies of other possible side effects of DES then got underway. Some studies examining the sons of DES users reported a higher-than-average incidence of various genital abnormalities, including abnormal sperm, and for both genders research showed abnormally high rates of severe depression, anxiety, and anorexia, along with other psychiatric disorders. Like thalidomide, here was another set of damages to the body's internal ecology brought about by a synthetic chemical. But DES stands apart from the thalidomide case most instructively.

Why did it take thirty years to find out the effects of this synthetic estrogen? During the same decade that DES was tested and approved for use, an experiment carried out at Northwestern University had shown that feeding pregnant rats additional estrogen leads to various abnormalities in the baby rats. This experiment hardly resonated, however, for as the authors of *Our Stolen Future* point out, "DES was just one of many new synthetic chemicals that promised to give us control over the forces of nature. With a mixture of hubris and naiveté, advocates of progress imagined a world with unlimited potential for the mastery of life itself."[52] We are fully familiar with this sensibility and know quite well how it deceives. Yet two other points must be made. First, it really was not before the problems arising with DES and, most significantly, not until the disclosures of Colborn's own work, that science, or at least a number of scientists, first came to realize that animal testing on endocrinological matters correlates much more closely with human results than does animal testing on questions of disease such as cancer. More important, however, is the proverbial nose on the face: the damaging effects simply did not appear, they were not to be seen, because it took them time to materialize. Unlike the immediate appearance of tragedy in the case of rat pups or thalidomide, there were no "DES babies." "This tragic and unintended experiment demonstrated that chemicals could cross the placenta, disrupt the development of the baby, and have serious effects that might not be evident until decades later. This was a previously unrecognized medical phenomenon: *delayed* long-term effects that did not emerge until the child reached puberty or anytime later in life."[53]

Ozone depletion is generally a case of ecological effects' being delayed by interactional realities of one kind or another. The temporal separation of cause and effect in that case is largely defined by the time it takes environmental interactions to work their causality. DES spotlights yet another dimension of the temporal separation of cause and effect. The story of global hormone-disruption told by Colborn and her colleagues places such interactional time within the life-time of two generations. And what we confront here in the body's possibly delaying effects for decades is, in principle at least, identical in its essentials with the problem confronting the prospect of our genetically engineering our organic future. We simply cannot foresee the potentially cascading effects of such organic intervention where developmental and intergenerational life-time governs the realm of appearance; where the body tells us what we need to know in its own time.

Making the right connections between cause and effect may be possible only after we have followed the long course of developmental processes. Colborn writes: "To screen for chemicals that can rob human potential before birth, it will be necessary to look for developmental effects across three generations—those individuals exposed only as adults and their children and grandchildren who inherit hand-me-down poisons."[54] Even close screening, however, offers no guarantee that causal connections can be made definitively. Indeed, the delay separating cause and effect may be so long as to render recognition of the connection somewhat a matter of luck. In this regard, while DES was judged a criminal and outlawed, Colborn and her colleagues wonder, nevertheless, if DES would have escaped detection for any of its criminal effects were it not for the eye-catching rarity of the cervical cancer giving medical science its first clue to the dangers of this synthetic chemical. They acknowledge being in a small minority in holding this thought, but they also are well aware of the fact that much of the knowledge we have of hormone-disrupting chemicals "has come to light largely through a series of accidental discoveries and surprises...."[55] Good fortune be blessed, the implications of the point here disturb nonetheless.

Screening a hormone-disrupting chemical over transgenerational time, while itself critical in cases such as these, guarantees nothing whatsoever in the way of knowledge because the chemical keeps widespread interactional company with many others like it. Isolating variables can be simply overwhelming when the transgenerational delay in the appearance of effects is considered along with the fact that these chemicals are to all intents and purposes ubiquitous. Bisphenol A, for example, is an estrogen mimic that has been widely employed for half a century, primarily in the manufacture of polycarbonate plastics used in kitchen appliances, baby bottles, compact discs, paints, and many food and beverage cans. The CDC has found levels of this estrogen mimic in human blood, placental tissue, amniotic fluid, breast milk, and, more broadly, in over 90 percent of human urine samples.[56] Even notorious outlaws such as PCBs, which were banned thirty years ago, are at large throughout the life-world. Given their chemical stability, which establishes longevity, and their widespread industrial employment for half a century prior to their being outlawed, these hormone-disrupting criminals are detectable in the body fat of virtually every animal alive. Our bodies may blend and mix all sorts of unwanted chemicals, and not simply those identified as hormone disrupting.

Hormone-disrupting chemicals themselves are but one class of chemicals, and since the Toxic Substances Control Act was enacted in 1976 over 80,000 chemicals have been produced.[57] We are swimming in a vast and extraordinarily complex chemical soup that can and does surpass our ken. Colborn and her colleagues develop this point fully:

> Even if damage is apparent and documented ... it will never be possible to establish a definitive cause-and-effect connection with contaminants in the environment. Although we know that every mother in the past half century has carried a load of synthetic chemicals and exposed her children in the womb, we do not know what combination of chemicals any individual child was exposed to, or at what levels, or whether he or she was hit during critical periods in their development when relatively low levels might have significant lifelong effects. This is a common and inescapable dilemma in trying to assess the delayed effects of environmental contamination. We also face the problem of having no genuine control group of unexposed individuals for comparative scientific studies. The contamination is ubiquitous.... In the real world, where humans and animals are exposed to contamination by dozens of chemicals that may be working jointly or sometimes in opposition to each other and where timing may be as important as dose, neat cause-and-effect links will remain elusive.... As we wrestle with the question of how much chemical contaminants are contributing to the trends and societal patterns we see—in breast cancer, prostate disease, infertility, and learning disabilities—it is important to keep one thing in mind...[:] The danger we face is not simply death and disease. By disrupting hormones and development, these synthetic chemicals may be changing who we become. They may be altering our destinies.[58]

Once again we face the problem of our being unable to specify the variables that may interact in combination or synergistically, but now the problem becomes magnified monumentally: the scale of the problem is one with the interactional manifold of the entire natural world into which we have been injecting tens of thousands of chemicals. Few of the new chemicals annually injected into the environment are tested as they should be. Yet, as criminal and foolhardy as this situation may be in itself, even the most properly conceived and precisely executed laboratory and field testing would fall far short of ever assessing what needs to be known for predictions of all the effects of our applied knowledge to have any real substance as foresight. Those who speak of controlling our evolution by genetic design would do well to give long and focused thought to the fact that nature, contaminated nature if you will, will perhaps have something very serious to say about who or what we shall become in the future. And in the meantime, the realm of appearance will be governed by lifetime.

We live in a state of uncertainty without any scientific means of determining what our own organic future holds amidst this wanton, interactional flow of ecological events processed through lifetime, our own and that of our progeny as well. Masters ready to chart our own destiny? The very question would be utterly laughable were the stakes not so serious. It is altogether possible that there is already underway some irreversible course of evolution to which we are contributing blindly. In this fundamental regard we are very much less like a species on the verge of assuming genetic control over its own evolutionary future than a species moving forward blindly, ignorantly, without any clearly defined map or reliable compass respecting its own organic future. The authors of *Our Stolen Future* put the point this way: "Judging from past experience, it may take a generation for the next nasty surprise to emerge.... If anything is certain, it is that we will be blindsided again.... We can screen chemicals for hazards that have already confronted us such as hormone disruption and ozone depletion, but the next nasty surprise will happen because we do not even know what questions to ask."[59] Not surprisingly, this problem of not knowing even what questions to ask recapitulates not only the critical point of much that we have discussed in this chapter on ecological affairs, but also much that critically concerned us in our discussion of genetic engineering in the preceding chapter.

None of this means that we should throw up our hands in absolute despair. Without assuming that we have things under control, the screening of chemicals at the very least should be undertaken across generations even though we may never establish any definitive knowledge. We certainly would not want to continue using a synthetic chemical if continued monitoring and tests implicated it in nasty eventualities that had not been even hinted at, much less foreseen, earlier. The problem of interactional, non–linear causality will remain, of course, and, with that in mind, perhaps serious thought should be given to following the more general advice of the authors of *Our Stolen Future*, who recommend the pursuit of alternatives to carbon-based chemicals wherever that might be feasible. But such prudential action presupposes a sensibility we have yet to demonstrate. In any event, the cases we have discussed in this chapter are meant dramatically to illuminate the limitations of scientific knowledge in much of our interaction with nature. We have been telling ecological detective stories featuring causalities eluding our experimental grasp. Such elusiveness moves the realm of appearance beyond scientific command, placing it in the domain of lifetime and turning scientists, or at least some of them, into observers who become scientific detective writers. Nature makes organic change as slippery as possible via indeterminate interactions, contingencies, and novelties, making life as a whole a terribly frustrating *object* for Baconian projects to grasp. Maker's knowledge has made over our world most impressively in modern time, but in the partially random fluidity and overwhelming complexity of the natural life-world, maker's knowledge actually does come up remarkably short.

Reaching this point would end our critical discussion of the illusion of mastery were it not for the fact that another call for mastery finds its voice at just this very juncture. Where organic nature and ecological interactions call into question any project of mastery defined by maker's knowledge, we find a number of people offering a new scientific perspective having its own instrumental program. This instrumental perspective first gained some purchase among global audiences in the 1960s when critical ecological consciousness was dawning. Calling attention to this new perspective in 1969, the futurist John McHale wrote, "We ... realize that man does not, in the end, master nature in the nineteenth century sense but collaborates within the natural world; his very existence depends upon an intricate balance of forces within which he is also an active agent."[60] McHale later makes it quite clear, however, that it is the Baconian conception of scientific control that is at issue, not mastery itself:

> The specific focus of our discussion centers around our capacity to control the enormous scale of our present global undertakings in a more positive and naturally advantageous manner and to avoid the dislocations and dangerous side effects of our technological and other growth patterns.... At that point, then, where men's affairs reach the scale of potential disruption of the global ecosystem, he invents precisely those conceptual and physical technologies that may enable him to deal with the magnitude of a complex planetary society.[61]

Planet-management thinking made its meaningful debut when serious environmental problems were beginning to surface in light of Rachel Carson's work. Substantively, the call for planet management does not discount straightforward control technologies of the sort associated with conventional science and industrial progress. Nor does it object in principle to the kind of mastery proposed by genetic engineers. This new perspective can embrace all such control capabilities as far as they go, but it claims to provide something fundamentally different in the way of scientific understanding and control—something going beyond mere maker's knowledge and altogether needed and timely vis-à-vis environmental problems of all sorts. As Erwin Laszlo spoke for this way of thinking while helping to promote it during the early years of environmental awareness: "Immersed in the immense hierarchies of the biosphere, we are nevertheless masters of our destiny, for we have enormous control capabilities.... The natural philosophy of the new developments in the sciences is a systems philosophy."[62] Here at the introduction of the planet-management perspective we have a new project of mastery in view: instrumental adaptation through systems control.

7

Planetary Thinking

Planet-management thinking is not cut from a single cloth. Viewed historically, it has changed its appearance as the crisis of the natural life-world has sharpened throughout these last few decades. The early perspective on planet management was the one offered by McHale and Laszlo which expressly identified instrumental adaptation with planet management and global controls in heraldic fashion. In the 1960s and 1970s the systems thinking associated with planet management was young and vigorous. Bucky Fuller was preaching his eccentric brand of global management in the idea of Spaceship Earth, while those following the organismic thinking of General System Theory, whose founder was the biologist Ludwig von Bertalanffy, promoted the idea that the new systems philosophy and systems science provided organizational handles on nature's complex interactional reality. During that time environmental problems were beginning to appear serious without being dire, at least to most people, and the systems thinkers and futurists promoting planet-management solutions to those problems could do so with happy eagerness. Evolution had prepared us to take control of our destiny, and the prospect of such mastery via design and management instruments was exciting.

This manner of thinking persisted into the closing decade of the twentieth century and is still evident in some quarters today. However, our worsening ecological straits— or, perhaps more importantly, our enlarging awareness of those straits in the last ten to fifteen years, and especially the scientific community's awareness of climate change — has altered the packaging and delivery of planet-management thinking. Although the instrumental contents are essentially unchanged, this thinking now has about it an urgency that has largely stripped control strategies of the grand talk of our species' taking its destiny in hand. Instrumental adaptation by planet management is now seen as more necessary than ever, but the bold, forceful expressions of mastery like those found in McHale and Laszlo have become fewer. While systems control by way of design and management science remains very much intact, the talk now is about sustainability and *stewardship*. Facing grave ecological threats of our own making, governing planetary interactions now means assuming life-sustaining responsibility for the future of our planet. The notion of our having a stewardship role on this planet follows directly from the biological sensibility marking most planet-management thinking right from the very start. At the same time, the idea of stewardship softens the hard surface of instrumental control; talk of such responsibility surely is more appealing than bombast about our vast control capabilities. Considering the historical development of thinking about planet management on the whole, then, while calls for planetary controls have gained stridency, the control being proposed has become more gentle-sounding.

We want to look at this development of planet-management thinking briefly in the first section to follow. Having considered this thinking generally in its historical development, we will subsequently turn to the control panel that systems thinking provides for the design and management of complex interactional wholes. Complex interactional wholes are the organizational reality that systems thinking addresses, and the meaning of control in systems understanding differs foundationally from the maker's knowledge we have treated throughout this study. Given our identification of mastery with maker's knowledge, it remains to be seen what bearing systems thinking may or may not have on the limitations of scientific knowledge already discussed in various contexts. In this chapter our critique of mastery takes a new turn as it follows the thinking of those who would have us believe that systems thinking and systems instruments can somehow make up for what maker's knowledge lacks. Ultimately, in the final section of this chapter we will return to storytelling in the land of sometimes and perhaps, focusing on the argument for global warming amidst the uncertainty of our time.

Instrumental Adaptation by Planet Management

Most systems specialists currently work as systems analysts in the management divisions of corporations or of municipal development agencies and the like. They have little if any professional concern for the matters concerning us in this study. So, too, systems thinkers who reflect theoretically or philosophically on nature are not necessarily bound in any way to claims or ambitions relating to planetary control. Our concern here is with planet-management thinking, which brings systems thinking to the fore in selective fashion. The discussion to follow focuses, therefore, not on systems thinking as such, but on that systems thinking speaking directly to the possibility and desirability of humankind's taking command of Earth to greater or lesser extent. Among those helping to lay the foundation for such thinking were many who sought to refound scientific enlightenment by unifying all the sciences, natural and social, upon new ground. That ground was General System Theory.

Those who founded the Society for General Systems Research in 1954 and established General System Theory exemplified the unification of the sciences: Bertalanffy, the leading figure, came from biology; Kenneth Boulding from economics; Anatol Rapoport from psychology; and Ralph Gerard from physiology.[1] Believing that reality is fundamentally and essentially organizational in its nature, these men and others following their lead attacked the reigning positivistic program of science that reduced everything to the model of physics. They directly criticized the positivist belief that all phenomena could be explained via causal analysis reducing everything to its smallest component elements. Such analytical reduction is inapplicable to those complex systems or higher-order organizational phenomena typified by living things, which constitute themselves interactionally and operate through non–linear causal processes. Life phenomena are self-organizing systems and call for a holistic systems understanding to address interactional complexity and non–linear processes. This biological bias of General System Theory did not, however, boil the program of unifying the sciences down to biology. The foundation proposed for unifying the sciences was, in fact, ingeniously non–partisan in its purely mathematical nature.[2]

Bertalanffy declared for General System Theory as a whole that a "unitary conception of the world may be based ... on the isomorphy of laws in different fields.... Speaking in what has been called the 'formal' mode, i.e., looking at the conceptual constructs of science, this means structural uniformities of the schemes we are applying."[3] Rapoport puts the same point as follows: "If it happens that several phenomena of widely different content can be described by the same mathematical model, both simplicity and generalization have been effected. Therein lies the power of *mathematical* general systems theory, based on the principles of mathematical isomorphisms, the completely rigorous version of analogy."[4] Isomorphism in this context pertains not to any direct correspondence between mathematical equations of knowledge and natural phenomena, but, rather, to a relationship between or among mathematical equations of knowledge considered in and of themselves. If laws pertaining to different phenomena have identical mathematical form, then these phenomena manifest the same organizational reality. This mathematical formalism of itself yields not phenomenal content, but the organizational structure of constituent properties. As the cookie-cutter may be considered apart from the cookies, so this isomorphism gives us to understand along with Laszlo, that "[n]ot what a thing is, what it is made of, or for what purposes it exists, defines it, but how it is organized."[5]

Isomorphism interests us no further, as we intend this idea mainly to identify at the ground level of the proposed unification of the sciences the fundamental abstraction of the idea of organization. The concept of system reproduces this abstraction definitionally. In Bertalanffy's spare conceptual definition, systems basically are "complexes of elements standing in interaction."[6] The reality of a system lies in the interactional process(es) of its constituent elements. Interaction also defines a system's relationship to its environment, the environment being abstractly identified as that which relates to the system but lies outside of it. "For a given system, the environment is the set of all objects a change in whose attributes affect the system and also those objects whose attributes are changed by the system."[7] What conceptually distinguishes the interactions internal to the system from those relating the system to its environment turns on the criterion of controllability, which is a matter that must await later discussion. Here amidst these vague and slippery abstractions, we must understand simply that systems undergo change through interactions that lead either to increasing organizational complexity and order or to increasing structural breakdown and disorder.

Lower-order systems consist entirely of physical elements subject to entropy. Systems thinkers call such organizations thermodynamically doomed to ineluctable disintegration *closed systems*. Subject to the second law of thermodynamics, the organization defining closed systems disappears in the irreversible breakdown of structural differentiation. Life phenomena are altogether different from the standpoint of system theory. Organisms constitute higher-order systems inasmuch as they exhibit increasing organizational differentiation, or negative entropy, both biologically, in their internal development, and historically, in the evolution of life as a whole. These systems are *open systems* in that they can process interactions with their environment to produce energy, both to offset entropy, as when a body repairs or heals itself, and, moreover, to attain greater organizational complexity. Bertalanffy writes:

> A living organism maintains itself in the state of highest organization, in a state of fantastic improbability. We can go even further: during differentiation an organism passes from states of lower to higher heterogeneity. This holds true for embryonic and postnatal development as well as for phylogenetic evolution from simple to higher forms of life; in both cases the transition is towards ... systems of higher order and organization.[8]

Open system is General System Theory's answer to the question "what is life?" Arriving at this answer via a conceptual path between physical reductionism and metaphysical teleology, general system thinking sees life's purpose as nothing other than the principle of *self-organization*, which individual organisms exhibit in their systemic nature. Organisms ceaselessly seek to survive, and they organizationally process their interactions with the environment toward that end. Negative entropy is not confined to the life of individual organisms, however, as organic nature generally realizes higher levels of self-organization in open systems of increasing complexity. Considered historically, as general systems theorists see life as a whole, self-organization means evolution, which leads to ever higher levels of organizational complexity in nature's organic development. As Bertalanffy tells us, "Life spirals laboriously upwards to higher and ever higher levels.... It passes into levels of higher differentiation and centralization.... It invents a highly developed nervous system and therewith pain. It adds to the primeval parts of this nervous system a brain which allows consciousness that by means of a world of symbols grants foresight and control of the future...."[9] What appears here, quite plainly and simply, is instrumental adaptation à la systems thinking.

From the systems standpoint, the possibility of controlling instrumental adaptation arrives exactly with the evolutionary emergence of systems understanding itself. Systems understanding brings to self-consciousness the self-organizing nature of life as a whole. Erich Jantsch writes in his *Design for Evolution*,

> [A]n evolutionary perspective of human and social life is more than just an interesting hypothesis—[it] is, indeed, part of the new emerging synthesis.... Man with all his powers of design ... and all his technology, is a manifestation of evolution as is all the rest of nature. But evolution is more than change, more than an array of interacting processes.... It is the self-organization of processes, this ordering of orders of change, which is coming more sharply into focus as evolutionary theory merges with dynamic general system theory.[10]

In systems understanding instrumental adaptation becomes one with the idea that humankind should become self-consciously self-organizing. Humankind should design and manage this planet Earth in all or most of its myriad systems, both artificial and natural. Such control may be conceived loosely in limited, decentralized fashion, or more strictly in authoritatively comprehensive terms. Instrumental adaptation à la systems thinking in any case highlights the need for human beings to design the future. Whatever the particular designs members of this early generation of planet managers may have had in mind, those calling for systems control were primarily concerned with normatively reorienting humankind's understanding towards a new way of thinking, towards a systems perspective. The planet-management thinking of the 1960s and 1970s was as much as anything else a call to embrace systems understanding. Hence, the heraldic sound of so much of that thinking.

At the height of such heraldry were heard voices riding a cresting wave of enthusiasm, a new wave, if you will. One representative voice belonged to Victor Ferkiss, whose *Technological Man: The Myth and the Reality* dramatized matters to a not inconsiderable readership. Calling for the birth of technological man, Ferkiss wrote: "Man is standing over the prostrate but still writhing figure of nature, trying to decide whether to kill it, make it his slave or free it to become his partner.... What will man do with nature and himself?... Technological man does not yet exist. His job is to invent not the future but first of all himself."[11] Technological man is a master whose mastery directly connects technology and evolution. "Technological man will be man in control of his own development within the context of a meaningful philosophy of the role of technology in human evolution.... [T]o control technology, to control the direction of human evolution, we must have some idea of where we are going and how far, else we will be mere passengers rather than the drivers of the chariot of evolution."[12] The drivers of the chariot of evolution unsurprisingly possess a systems perspective on things, but such understanding must gain much wider currency for technological man to be born as such. In Ferkiss' words:

> A basic element of this new philosophy is what might be called the *new naturalism*, which asserts that man is in fact part of nature rather than something apart from it.... Closely related to the new naturalism is the *new holism*, that is, the realization of how connected everything is.... The image of the mechanical universe must give way to the idea of process. The basic concepts of process and system imply a recognition that no part is meaningful outside the whole.... But this whole, the universal as well as the social, is a new kind of whole, determined not from outside but from within. For another element in the new world outlook is the *new immanentism*.... However physicists may look upon the development of the physical universe as a whole, the world of living things is somehow different. Nature here works another way, life is *antientropic*.... Life exists within systems. And systems create themselves.[13]

Self-organization at the highest level embraces many systems, artificial institutional systems (social, economic, political, etc.) as well as natural interactions. Those calling for a reorientation of consciousness as often as not placed emphasis on institutional systems, since those systems would necessarily require change to accommodate planet management. Yet even for the early generation of planet managers, real-world problems in themselves called for a systems approach. As Bertalanffy declared without any fanfare, "Systems of many levels ask for scientific controls; ecosystems ... the grave problems appearing in socio-economic systems [etc.] ... [and] there can be no dispute that these are essentially 'system' problems, that is, problems of interrelations of a great number of 'variables.'"[14] Or, as Laszlo put the point in a popular image of the time, "The values of today are inadequate to handle the constraints of tomorrow; but today's values, fortunately, are changeable. Soon we shall be living in a spaceship almost as vulnerable and in need of purposeful guidance as today's airplanes."[15]

Not all planet-management thinking in this early generation cited General System Theory, but the thrust of much of that thinking came from the promise of the work being reported in the General Systems Yearbook and related journals. That promise faded as the century wore down and General System Theory's aim to refound scientific enlightenment via isomorphism fell short. General system theorists still do carry on, and the scientific study of complex systems and many of the themes of General System Theory would later resonate loudly in the related quarters of chaos theory and, more especially, in the work of the Santa Fe Institute and complexity theory in general.[16] Our concern is with planet-management thinking, however, not complexity theory per se, and that thinking in recent years generally has abandoned the grand promise-making. Metaphorical references to the chariot of evolution or to spaceship Earth have become fewer, appearing only occasionally, as planet-management thinking by and large has become more sober. Nothing has changed in the claim to control, the control panel itself having become merely more sophisticated with the evolution of super-computers. What has changed is a growing sense of urgency concerning the environmental problems endangering our well-being. Climate change, calamitous environmental degradation, and the like have impressed upon planet managers the urgent need for control, but in the place of chariots and spaceships we now mainly hear talk of sustainability and stewardship.

The focus on sustainability concerns how we as a species will have to adapt our institutional practices, broadly considered, and change our "ecological footprint" to establish and maintain acceptable living conditions. Sustainability loosely refers to our metabolic relationship to Earth, and issues of sustainability are myriad, involving everything from increasing population and dwindling natural resources to matters of global warming, desertification, species' extinction, environmental pollution and the like. Sustainability refers, in short, to all global interactions relevant to the planet's life-support systems. Given such breadth, calls for sustainability come from many people having widely divergent perspectives, concerns, and interests. Talk of sustainability may respect local community issues or individual household choices, or it may concern global problems having very little to do with anything apart from hard choices societies are going to have to make to sustain some level of desirable living conditions in the future. Our critical concern, therefore, is not with sustainability itself, but with where the interest in sustainability intersects with planet-management thinking. Crutzen's notion of the Anthropocene has gained currency perhaps partly because it pointed directly to that intersection.

Speaking of this new geological age, Crutzen wrote that, barring some global catastrophe like a meteor's impact or a world war, "mankind will remain a major environmental force for many millennia. A daunting task lies ahead for scientists and engineers to guide society towards environmentally sustainable management during the era of the Anthropocene. This will require appropriate human behavior on all scales, and may well involve internationally accepted, large-scale geo-engineering projects, for instance to 'optimize' climate,'" and as he concludes, "At this stage, however, we are still largely treading on *terra incognita.*"[17] Wanting to map that territory are planetary thinkers who link problems of sustainability to systems understanding and strategies of adaptive control. One group of authors thinking generally along this line warn that "[i]n the absence of effective governance institutions at the appropriate scale, natural resources and the environment are in peril from increasing human population, consumption, and deployment of advanced technologies for resource use, all of which have reached unprecedented scales."[18] Having addressed requirements and strategies for "adaptive governance in complex systems," the authors then conclude with this: "Is it possible to govern such critical commons as the oceans and the climate? We remain guardedly optimistic.... Systematic multidisciplinary research has ... shown that a wide diversity of adaptive governance systems have been effective stewards of many resources.... As the human footprint on Earth enlarges, humanity is challenged to develop and deploy understanding of broad-scale commons governance quickly enough to avoid the broad-scale tragedies that will otherwise ensue."[19]

Another group of authors is concerned that our understanding is underdeveloped in the four disciplines they see most critical to problems of sustainability. Demography, ecology, economics, and epidemiology "share a limited ability to appreciate that the fate of human populations depends on the biosphere's capacity to provide a continued flow of goods and services.... Many disciplines still apply worldviews that predate current understanding of complex system dynamics...."[20] Speaking more particularly of ecology, these authors note that this discipline increasingly does appreciate the interplay between the natural world and human systems, but they remain somewhat critical of how far such sensibility has advanced: "[V]arious conceptual and theoretical frameworks in ecology still disregard the connection of ecosystems to the human species. More integrated views from landscape ecology and systems approaches, and the greater appreciation of complex systems, critical thresholds, and the possibilities of state changes, are attracting attention."[21] They therefore call for an integrative *sustainability science* that "is an emerging science that seeks to understand the fundamental interactions between nature and society. It encompasses the interaction of global processes with the ecological and social characteristics of particular places and sectors. It addresses such issues as the behavior of complex self-organizing systems as well as the responses, some irreversible, of the nature-society system to multiple and interacting stresses."[22]

Talk about complex self-organizing systems or about adaptive governance in complex systems is, of course, fully in keeping with the language used by the first generation of planet managers. So, too, the abstraction of that language from the ground floor of real-world ecological interactions remains intact, and we will have much to say about this abstraction later in this chapter. More immediately, however, this current thinking does not sound quite the same as that of the earlier generation. "Governance systems" and "sustainability science" have replaced the bold, hard-edged conquest metaphors; they have a softer feel and seductively do ring more sensibly in our ears. But occasionally the newer interest in sustainability combines with the older, bold enthusiasm for instrumental adaptation, and we get an interesting mixture of urgent need with illustrious possibility. Such a mixture is the work on planet management edited by Norman Myers and Jennifer Kent, now in its third edition. In their introduction to *The New Atlas of Planet Management*, Myers and Kent write the following: "Techno-wizards we may be, yet we don't really sense what we're doing — how we are changing our planet, our lifestyles, our aspirations to an extent that should leave us gasping with the sheer onrush of change.... Those daunted ... should remember that the choice is not between change and no change. Rather, it is between change that we choose for our future and change that we suffer as a result of our past. Shall we choose to choose?"[23] This question is entirely rhetorical since for roughly "four billion years, the experiments, the increasing diversity, and complexity [of evolution] continued until, as the life-support systems of our planet stabilized, a still higher order of complexity emerged — intelligence and conscious awareness.... Evolution has also equipped us to create our own form of planetary ecosystem."[24]

Myers and Kent are quite confident of our capabilities, and in their concluding chapter, which is entitled "Management," we learn that issues of sustainability really involve little more than our will and a new ethos. They tell us this in language reminiscent of the 1960s and 1970s: "We are a privileged generation, to be living at a great turning point in the story of Gaia and humanity.... After setting ourselves apart from nature, we must become a part of nature: truce, treaty, and reconciliation — and so become, truly, humankind. This is a challenge indeed, one that requires a sweeping revolution in how we act, think, and feel; in how we use our technology; and above all, in how we manage ourselves ... [;]" in short: "[t]he transition to a sustainable society is eminently possible. We have the tools but we lack the will."[25] One apparently unimportant question is how we are to identify our role in our upcoming interaction with nature. "There is sharpening debate over what should be the role of humans in relationship to planet Earth. Rather than 'managers,' how about 'stewards' or 'caretakers?'"[26] But whatever title we ultimately give ourselves, our capabilities are not in question. "We now have the skills to derive benefit from natural systems without destroying their ability to renew themselves. We can live off the biosphere's income rather than depleting its capital. We can cause forests to spring up on barren lands, safeguard species from extinction, renew the soil, reverse the pollution of oceans and atmosphere, harness the renewable power of the sun, and minimize the consumption of resources."[27]

Reading of such possibilities, some of us may be tempted to jump up and applaud this watershed moment as we answer the call of the Anthropocene and, with full evolutionary entitlement, take responsibility for this planet and our future — via management, caretaking, or stewardship, take your pick. Most likely it is this feel of 1960s exuberance in the face of late-twentieth-century planetary threats that led Bill McKibben to comment on an earlier edition of Myers' *Atlas*, that it cloaks a real interest in control "in a filmy veil of ecological New Age thinking."[28] McKibben, one of the very best among those who think and write about our relationship to this planet, is unconcerned with the intellectual background to this "macromanagement" thinking, while we, on the other hand, address Myers' work for the very reason that it does, in fact, draw together the grand confidence and expectations of earlier planet-management thinking with the urgent concern for sustainability marking much contemporary environmental thinking. With or without a filmy veil, this thinking means instrumental adaptation via managerial systems control.

The Irony of Systems Control

Talk of sustainable solutions may ignore instrumental knowledge altogether, or it may only vaguely refer to the kinds of disciplinary knowledge needed. In any case, there is a technology, or, better, a combination of instrumental practices, that belongs to planet management. This technology consists essentially in organizational design and management instruments along with the computer wherewithal to make those instruments effective. This instrumental control capability emerged alongside the systems understanding we have discussed, but it was not born within general system science, nor even from a single source. Its parents were information theory, especially under the influence of Norbert Wiener's cybernetics, and systems analysis (or operations research), which entailed organizational design and management strategies developed initially in military and industrial bureaucracies. We will refer to this control capability simply as systems technology in what follows.

Organizations public and private, big and small, widely employ systems technology in local contexts. Yet even in local contexts where systems thinking has no concern whatsoever with either planet management or nature's organic evolution, systems technology directly connects with the idea of instrumental adaptation. Organizations employing the design and management practices of systems technology understand themselves to be self-organizing. They seek via that technology to organize their own evolutionary change by controlling their organization's adaptation to environmental events and circumstances arising interactionally as they move into the future. In his "Human System Design and the Management of Change," Raymond Studer puts this operational meaning of instrumental adaptation in prescriptive terms: "In approaching the future, our design objectives should correspond more directly to the variable nature of open systems ... [as] a vehicle for moving from the present to an uncertain future in a controlled manner."[29] At the practical level of systems technology, instrumental adaptation *qua* self-organization means managing uncertainty and controlling the changes an organization or system may undergo in its future interactions with the environment.

Critical to this practical understanding of self-organization is the distinction between system and environment, which conceptually turns on the criterion of controllability. The criterion of controllability was the focus of those military and industrial organizations that originally developed systems analysis to manage and minimize internal problems arising from interactions with the surrounding world of clients, customers, and competitors, who were all independent actors and therefore beyond the control of the organization. In line with such grassroots situations is the conceptual demarcation given for all systems and environments by Dickerson and Robertshaw in their *Planning and Design*. They write:

> To determine whether something is part of the system or part of the system's environment, ask these two questions: "Will it affect the function of the system?" "Can its action on the system be controlled?" If the answer to both questions is "Yes," then the thing is part of the system. If the answer to the first question is yes and the answer to the second question is no, then the thing is part of the environment of the system.[30]

Turning this understanding around and simplifying the point, a system controls its own internal interactions but not those belonging to the environments with which it interacts. The question of whether or not something is controllable is, therefore, purely organizational: is it internal or external to the system?

This question of control is organizationally fluid inasmuch as the boundaries theoretically defining a system are very much a matter of perspective. Hall and Fagen pose the question of whether or not elements interacting with a system might not be considered part of that system, to which they respond: "The answer is by no means definite [since] ultimately it depends on ... the one who is studying the particular universe as to which of the possible configurations of objects is to be taken as the system."[31] In other words, what constitutes a (controllable) system and an (uncontrollable) environment changes from one organizational perspective to the next. Consider the instructor whose career path takes her from the classroom, through the office of department chair, to the office of division head. What once had been uncontrollable environments lying beyond the system of the classroom, have become, in time, manageable systems falling under the supervisory authority of the division head. This perspective-based nature of systems understanding provides systems thinkers and their design technologies wonderfully flexible employment; they can go almost anywhere. In this same vein, of course, the reach of control in systems thinking will go as far and assume proportions as large as the systems thinking will have it. Planet managers think very big: this planet is their system.[32]

This abstraction of the organizational understanding of reality from phenomenal content runs throughout the conceptual domain of systems theory. Therein lie both the strengths and weaknesses of the systems perspective. Organizational design and management technologies obtain power and value in their capacity to handle any and all kinds of organizational entities, including immensely complex wholes possessing large numbers of interacting elements. Be the application of systems science large or small, however, the technology employed in systems control is essentially the same. It consists primarily of computer experimentation and information feedback systems. Experimentally handling dynamic interactions even in relatively small or limited systems is a complex task requiring information processing, and on a planetary scale super-computers far exceeding the capabilities of office machines are necessary. In systems science computers are, in any case, the *sine qua non* of all serious experimentation, and the power of computers to process vast amounts of information provides systems science whatever credentials it possesses as an experimental science. Scientific experimentation in this context means, however, something quite different from what we have discussed in the traditional context of conventional science and maker's knowledge.[33]

Systems experimentation entails designing a computer program that models the behavior, or the interactional dynamics, of the real organizational system under study. In the computer model of a system the realm of appearance therefore stands as an abstract analogy to the reality the design wants to capture. The entire process of experimentation is itself an abstraction from the real world as the model is developed and checked against reality by way of computer programs. Systems experimentation operates by means of computer simulation, which is "a numerical technique for conducting experiments on a digital computer; this technique involves certain types of mathematical and logical models that describe the behavior of ... systems (or some component thereof) over periods of time."[34] Simulation operationalizes the design; it "runs" the interactional dynamics of the model through temporal projections built into the program.

The systems experimenter aims to optimize the simulation by first studying the real system. Knowing something about the history of the real system is critical to designing a computer model of it; historical knowledge helps the experimenter to select the parameters and establish the variables important to the behavior being investigated. Selection of the parameters, or "parameterizing" in the jarring systems vernacular, means fixing—making fixed—those analytical dimensions that are to frame the interactional elements of the system. Parameters identify the constants that shape the design. Parameters also "gloss" the real world, as it were, generalizing or averaging where complete knowledge goes wanting. Identifying parameters in any case expresses the experimenter's own selective attention to the real world, or to that portion of the real world being modeled. Uncertainty attaches to that selection in the same way the experimenter's perspective and knowledge inevitably fall short of that possessed by the Laplacean seer. The uncertainty occasioned by such subjective selection is, however, not the only uncertainty in the experimentation.

A different kind of uncertainty is that which the experimenter him- or herself designs as part of the model. The experimenter builds this uncertainty into the simulation in recognition of the open-ended interactional dimension of complex non–linear systems. "The uncertainty that commonly attends the behavior of systems ... can be taken account of in terms of the theory of *stochastic processes*. In the stochastic model of a system the state of the system at time *t* is defined in terms of the probability of distributions of the alternative values that any physical variable of the system may assume at the time."[35] Stochastic processing provides for variable futures of the system and channels the model's interactional dynamics along probabilistic lines defining uncertainty in experimental terms. Unlike the uncertainty abiding subjectively in the choice of parameters, this uncertainty obtains some mathematical specificity, some "objectivity," as it were, in the abstract realm of the computer simulation.

Once the parameters and numerical values relating to interactional elements have been designed, and the stochastic measures have been put in place, the experimenter is ready to proceed. He or she can process the model's dynamics by running the simulation:

> Simulation is essentially a working analogy.... Simulation involves the construction of a working ... model presenting similarity of properties or relationships with the ... system under study. In this manner we can preoperate a system without actually having a physical device to work with, and we can predecide on the optimization of its characteristics.... A simulation study begins with the development of this custom-made model and continues with its processing or "operation" in order to determine the behavior of the systems....[36]

Preoperating and *predeciding* means comparing the model's dynamics to those of the real system and changing numerical values or allowing parametric adjustments to achieve as close an analogical match as possible. Experimentally preoperating and predeciding a system further allows the experimenter to test different policy alternatives and goal strategies. Varying the simulation along alternative interactional paths permits the experimenter to anticipate different possible futures. He or she can test-run different organizational choices and policy paths and judge from among these various scenarios the one that is most likely to achieve system objectives while best managing uncertainty. Ultimately, of course, actual choices are meant to be implemented upon this experimental basis. And those choices, once implemented, in themselves manifest self-organization on the practical level of instrumental design and organizational management. Put the other way around, self-organization in human systems, big or small, means managing uncertainty and controlling organizational change via policies and strategies worked out through computer design and experimentation.

The key to simulation's managing uncertainty and controlling change lies in simulation's temporalizing the realm of appearance in the experiment. The computer simulates real time when the model runs. Dimitris Chorafas puts the point of simulation this way in his *Systems and Simulation*: "We use it as a tool in experimentation, as an extension of the scientists' penetrating power in problems of complexity.... Most important of all we use it as a lever in compressing time."[37] Or, as George Morgenthaler writes most bluntly on the power of systems experimentation, "Simulation gives us a control over time."[38] The great strength of systems analysis lies in this ability to handle time apart from the real world; systems experimentation enables us to temporalize by computer design the conditions of observation for dynamic, non-linear systems possessing interactional complexity. But the fundamental shortcoming of this instrumental means of "grasping" reality becomes apparent here as well. Operationalizing time by instrumental means suffers the selfsame abstraction as the computer model's distance from the real world.

While there is no good experimental alternative to computer simulation where complex interactional dynamics are concerned, the basically "unreal" nature of the simulation must be always kept in mind. Maisel and Gnugnoli put it plainly: "A computer simulation exists only in the set of computer programs [italics deleted]...."[39] In neither design nor operation does the simulated system actually engage reality in any immediate way. As the very term *simulation* suggests, the manipulation of parameters and variables *feigns* reality. The control of time and the handling of uncertainty likewise are feigned, and systems practice must account for this shortcoming and offset it. Systems thinking somehow must put experimentation in contact with the phenomenal world, and systems practice accomplishes this by embedding experimental design in the ongoing process of feedback control. Feedback control is fundamentally essential to systems control in general.

The feedback control we have in mind here loosely conforms to the control theory Norbert Wiener articulated in the idea of cybernetics.[40] Apart from distinctions between open and closed systems, and positive and negative feedback, feedback here refers very simply and generally to a system's continuing incorporation and use of information about its performance in its self-organizing practices. Since simulation only feigns reality, once an organizational strategy is implemented, information about the interactions that actually do occur and do effect change need to be continually fed back into the organization's operating system. Most critical in this respect is information regarding developing discrepancies between what is expected to occur on the basis of simulated reality and what does occur in fact. Learning about such discrepancies is significant if a system's performance is to remain robust, healthy, or "true." Feedback thus enables an organization or system to undertake action to alter its performance. Wiener likened this feedback-based correction of performance in a machine system (what is termed negative feedback in cases where corrections return performance to a given path) to a helmsman's steering his craft through the continuing adjustment of his actions. The term *cybernetics* was born of this steering analogy and the concomitant idea of feedback "learning."

Systems thinker C. West Churchman has noted that "Wiener and his followers developed a theory of cybernetics which has mainly been applied to the design of machinery. But it is only natural for the management scientists to attempt to apply the theory to the management control of large organizations."[41] It certainly makes sense for systems analysts to think in terms of feedback when considering the real-world limitations of simulation. Feedback processing very directly connects systems experimentation with reality's refusal to follow the simulation model. Feedback makes simulation no less abstract in its nature, but it does provide ongoing information about the real world that can be fed back into later, updated experimentation. As information processing becomes continuous in feedback functions, simulations can be redesigned as needed, and organizational choices can be continually undertaken anew through the serialization of experimentation in real time. Again, we listen to Churchman:

> Once a plan is implemented, there must be control, which includes feedback of information about the operation of the plan and change of plan when needed.... [I]t is essential that the planning function be designed so that, when a plan is implemented, it is possible to feed back ... information about what has occurred and to lay out in general terms the steps that must be taken for change.... One can see that this phase of planning does require recapitulation of all the steps, so that, as additional information pours in, correct change can occur. A change of plans is in effect a new plan and must be based on examination of each of the preceding steps, even the step of organizing for planning.[42]

Technologies of organizational design and management in this way identify systems control with an information processing system: simulation embodies such processing experimentally, and information feedback functions to control such processing in real time. Systems technology is thus an intellectual technology in that it consists in the handling and organization of information. Computers materially embody this process, but only within the framework of the designs and programs fed into them.

The abstraction we have spotlighted in systems thinking throughout our discussion in this chapter reappears here in the belief that control is by and large an issue of processing and managing information. The most serious shortcoming of systems experimentation springs from the essential abstraction of systems simulation — the abstraction of both the simulated model and the control of time—from the real world. Hence, planners build feedback into the process of control and "input" the continuous flow of information about the real world into ongoing simulation and continuing organizational adaptation. Feedback in this way operates in the gap separating simulation and reality; it functionally reacts to reality's disagreement with human design. Feedback control thus shifts the overall meaning of control from the experimental context of some simulated model to the time-serialization of experimentation itself. Experimentation and planning become processual, and in this time-serialized processing of information the very nature and meaning of systems *control* turns back upon itself in thoroughly ironic fashion.

Conceived as a steering function, feedback control temporalizes the conditions of observation, which means that real time itself now governs the realm of appearance in the real world. Real time governs the realm of appearance in the same measure that experimental simulation fails to predict or control the interactions causally effecting the actual development of the real system. Far from offsetting such failure, feedback's temporalization of the conditions of observation exhibits this inability to control things experimentally in the very meaning of feedback *control*. The irony here is huge, of course, for what passes for control in systems understanding actually manifests the absence of control. Quite literally, that is, feedback control is the attempt to manage—to keep within bounds—our *inability* to control complex, interactional phenomena experimentally through simulation and managerial design. For all of its sophistication, and notwithstanding the fact that incoming information will inform further simulation and planning, feedback control's temporalizing the conditions of observation ultimately amounts to monitoring, plain and simple.

Like the proverbial nose, this irony stands out as obvious and elementary once we see it. Systems thinking tends, however, to veil its irony in abstraction. On the face of it, systems thinkers seem to be telling us the very points we have used in criticism of their understanding of control. Systems practitioners are very clear that organizations employ systems technology expressly in order to manage the uncertainty arising from interactions lying beyond the control of the organization. Systems analysis is all about planning for an uncertain future in a complex and dynamically changing world. In other words systems thinking quite openly admits to everything here deemed ironic, minus the irony, of course. Are we not, then, really making much ado about nothing and, in the end, attacking a straw man?

There certainly is merit to this rejoinder so long as we imagine such a defense being argued by systems thinkers bound to local organizations having to wrestle with intractable difficulties lying beyond their systemic reach. We have no quarrel with systems thinking in such situations where the applications of systems technology are inconsequential to the issues of mastery occupying us in this study. However, it is just in respect of such issues that the abstraction and irony of systems control become critically telling. The problem with abstraction and irony in systems understanding appears strikingly just with those systems thinkers who present us with proposals to control our planet's future via design and management technoscience—with those systems thinkers, that is, who think big, as planet managers do. As we approach the planetary level, the concept of a system's controlling its interactions reaches critical mass, as it were. Now the troposphere and everything below it fall within the domain of the controllable, and one wonders what might constitute the intractable environment: the solar system perhaps? or the galaxy? For some visionaries, even these realms will come within our reach in time.[43] In any case, once this planet Earth begins to appear subject to design and management technologies, with or without the spaceship metaphor we encounter proposals for self-organization that suffer the abstraction of systems thinking, abstraction completely missing the irony of systems control.

Planet management thinking can soar in its designing vision as far and as high as human imagination may carry it, but it flies over the phenomenal ground of biological and ecological causality as if mastery of nature has some formula superior to the Baconian. Be the causality simple and linear or complex and elaborately interactional, mastery of nature has no instrumental formula other than maker's knowledge. We should not do without systems technology; indeed, we very much need it. But the problems besetting our relationship to nature hardly reduce to organizational terms having managerial solutions. Indeed, what on earth does systems control or the management of uncertainty mean if some of the chemicals wantonly pouring into the real living environment are currently conspiring in their causality, while holding on to it through transgenerational time, to effect a massive reproductive calamity upon our grandchildren? What would planet management or planetary system controls have meant throughout the decades of CFC use? or during the time of Gaeton Dugas' sexual exploits? In the face of such potentially devastating eventualities as ozone depletion or the spread of HIV, systems control and the management of uncertainty would amount to nothing more than horrific news following upon decades of monitoring. Hence, as planet-management thinking flies in abstraction above the Baconian ground of maker's knowledge — where, once again, time becomes telling where mastery goes wanting — mastery à la planetary management unwittingly speaks directly to our point. For after all is said and done, what really is the point of feedback control, where does it all lead, if not storytelling? Storytelling without end.

Planet Modeling and the World Ahead

Here we take leave of the abstraction to which the conceptual nature of systems understanding condemned our discussion in the previous section. Much remains to be seen in planet modeling, but we want now to shift the discussion away from conceptual-theoretical matters to empirical instances of modeling, beginning with a few remarks about the Club of Rome's work over these past thirty-five years and then moving on to a discussion of the argument for global warming.

Planet modeling first gained widespread notice when the Club of Rome published *The Limits to Growth* in 1972. Initially consisting of two and a half dozen scientists, industrialists, and other concerned international figures, the Club of Rome saw global peril looming in the interactional matrix of unchecked population growth, unmitigated exploitation of non–renewable natural resources, humankind's expanding pollution of the environment, and a host of other related threats to civilization. The Club engaged a group of MIT scientists to study and report on this matrix, variously called the "world problematique" and the "*problematic humaine*," and to offer alternative futures, at least in broad outline. Based on a world simulation model developed by the MIT team, *The Limits to Growth* was that report. Two years later, in 1974, the Club issued a second report under different authorship. *Mankind at the Turning Point* provided an updated and enlarged version of the "problematique" and offered more specific scenarios on what humankind might choose its future to be. Thus began an ongoing project that produced a third report in 1993 and then, in 2004, a work entitled *Limits to Growth: The 30-Year Update*, by three of the original MIT authors. Guiding this project throughout has been an unyielding belief that civilization ultimately is bound to break down and collapse if it continues on its present course. We must seize the day before it is too late to plan for sense and sustainability.

The 1974 report proffered its simulation not as a "predictor" but, rather, as a planning instrument for the user to "assess the consequences of implementing his vision of the future."[44] Although this distinction between predicting and planning may be too gross to be entirely persuasive, the Club's envisioning of possible futures has, in any event, been open-ended rather than deterministic. The authors of the 2004 *Update* put their large intention this way: "If we anticipate these trends, we may exert some rational control over them, selecting the best of the options available to us. If we ignore them, then the natural systems will choose an outcome without regards to human welfare."[45] In envisioning better futures, Club of Rome thinking also has generally called on values larger than science and rationality to inform the transition to a better, more rationally controlled human world. The latest report speaks of love and truth-telling, among other value commitments, as important to our altering our ecological footprint and bringing about planned sustainability in the place of irrational and non–sustainable social and environmental practices. Such sensibility seems clearly to suffer some of the ecological New Age thinking of which McKibben spoke in a different context, but these "soft" values only briefly relieve thinking that is otherwise instrumentally hard edged. The revolutionary thrust of this work is systems thinking through and through.

Calling for a revolution in perspective and practice does not mean running to the barricades, super-computers in tow. As these system changers argue in the 2004 report, "changing structure means changing *the feedback structure, the information links* in a system...."[46] Whatever else revolution might entail in the way we see and do things, "information is the key to transformation."[47] At present we lack the systems controls necessary to make possible rational planning for sustainability; everything is essentially dysfunctional. "Any population-economy-environment system that has feedback delays and slow physical responses; that has thresholds and erosive mechanisms; and that grows rapidly is inherently *unmanageable*."[48] Revolution thus means changing the information processing networks so as to make the planetary system manageable. Following upon their structural analysis of the modeled world system, these "systems dynamists," as they call themselves, offer seven steps in their guideline towards a rational, sustainable world.[49] The first three of these are most telling from our perspective: extend the planning horizon, improve the signals, and speed up response times. Elaborating this last point, these would-be planet managers write this:

> Look actively for signals that indicate when the environment or society is stressed. Decide in advance what to do if problems appear (if possible, forecast them before they appear) and have in place the institutional and technical arrangements necessary to act effectively. Educate for flexibility and creativity, for critical thinking and the ability to redesign both physical and social systems. Computer modeling can help with this step, but equally important would be general education in systems thinking.[50]

Systems thinking and systems control can provide the critical answer to our current plight. Not conquest in the old sense, but control à la the systems mode of design and management. Yet here, once again, we see the stunning irony in the abstract equation of control with information processing and management. How can control managers possibly decide in advance about problems that may not even be thinkable until they appear in the life-time of organic and ecological phenomena? What does the presumed ability to redesign physical and social systems amount to, if we are, potentially at least, subject to events or developments that might effectively overwhelm all such systems? Most generally, where does education in systems thinking along these lines lead but to a perspective on nature that promotes the sense of our being both collaborative and in charge? Thinking of this sort expresses belief in mastery, systems style. But planet modeling needs not be bound to ambitions or illusions of control. Such modeling may simply try to get some fix on the future amidst present uncertainty.

Just how uncertain we are about the world surrounding us and the possible future(s) facing us, has been nowhere more evident than in discussions of global warming and in the planet modeling experimentally investigating questions of climate change. Indeed, *uncertainty* has been the key word in those discussions, most notably for those scientists and organized interests who have for various reasons opposed the argument for global warming. Pointing to the lack of certainty as the Achilles heel of the argument for global warming has been the attack strategy of those various corporations and industries, as well as of political conservatives, who for various reasons fear the ramifications of governmental policies predicated upon global warming. This strategy stood stark naked in a leaked 1990 White House memorandum revealing President George H.W. Bush's administration's considered approach to the argument for global warming: the best way of subverting the argument for warming was to highlight "the many uncertainties."[51] But it would be absolutely wrong to write off uncertainty as if it were merely the ploy of those organized interests who feel threatened by the prospect of developing alternative fuels or of increasing governmental regulation and planning. Respectable scientists also have taken such a position. Uncertainty, in fact, is to some degree ineradicably part of the very argument that global warming is underway.

The argument for global warming fundamentally integrates three points. These three points more or less define the scientific consensus taking shape around the idea of global warming sometime near the turn of the millennium. At the heart of the argument is the knowledge claim that Earth's climate is warming at an unprecedented rate that will increase global mean temperature anywhere between 1.5 degrees and 4.5 degrees Centigrade (and on the outside, perhaps as much as 8 to 11 degrees Centigrade) during the twenty-first century.[52] Integral to this claim is the second, which is that anthropogenic causes are the driving force behind this increase, the most crucial of these being greenhouse gases like carbon dioxide and methane, which trap thermal radiation in the troposphere. And, given these two points, the third is that climate change will effect a host of ecological consequences which, depending upon whom one speaks to or reads, will be somewhere between the extremely undesirable and the dire.[53] Oceans will rise, salinizing coastal aquifers and overwhelming coastal settlements everywhere; biodiversity will suffer terribly as untold numbers of species unable to adapt to rapid change go extinct; species migration will spread and alter disease patterns, likely giving rise to new pathogenic threats as well; the frequency and potency of extreme weather events will increase; and human populations the world over will be severely challenged to accommodate themselves in some manner of rapid adaptation. Many other ill consequences can be found in the literature, but let this short list suffice. The pressing problem across the board, in any case, is generally thought to be the unprecedented rapidity of the projected change, which amounts to an ecological jolt the likes of which human beings have not experienced in their recorded history. If the argument for global warming is right, and we fully believe that it is, there is real likelihood that our progeny are going to face challenges for which the past may prove a terribly limited and somewhat unreliable guide.

Those making the case for global warming have about them an understandable urgency, for they want human action and public policy immediately to confront issues like energy use, population growth, and desertification, which all bear directly on problems of warming. Their argument has left them somewhat vulnerable to the opposition, however, precisely because their case lacks the firm Laplacean ground of predictive knowledge offering certainty. Long-time scientific opponent Fred Singer, for example, argues that before any binding restraints should be put on carbon emissions, the consensus must first and foremost conclusively demonstrate that "the greenhouse gases are certain to raise previous global temperatures significantly higher than they rose during the previous natural climate warming cycle...."[54] Singer is being disingenuous; he knows very well that his demand cannot be met. Reaching consensus among a majority within the scientific community was itself an accomplishment and took some time precisely because no certain demonstration of global warming is available. Obtaining consensus required many scientists' overcoming not only their natural psychological desire for certainty but also their tradition, which demands that knowledge claims be experimentally tested and demonstrably proved. Whatever *strong probability* or *very real likelihood* might mean in the way of belief, it departs fundamentally from that tradition. Yet *very real likelihood* must suffice for those who argue from *reasonable conviction* that global warming is real and that it threatens our well-being. Spencer Weart speaks for himself but with illumination where he writes the following:

> My training as a physicist and historian of science has given me some feeling for where scientific claims are reliable and where they are shaky. Of course climate science is full of uncertainties, and nobody claims to know exactly what the climate will do. That very uncertainty is part of what, I am confident, is known beyond doubt: our planet's climate can change, tremendously and unpredictably. Beyond that we can conclude (with the IPCC) that it is *very likely* that significant global warming is coming in our lifetimes. This surely brings a *likelihood* of harm, widespread and grave. The few who contest these *facts* are either ignorant or so committed to their viewpoint that they will seize on any excuse to deny the danger [second and third italics added].[55]

In one breath Weart refers to global warming as being very likely, and in the next he speaks of this likelihood as a fact. Exactly how likelihood and factuality manage to shake hands here apart from mathematical factuality is not altogether clear except in Weart's own "feeling for where scientific claims are reliable and where they are shaky." We do not dispute the legitimacy of that feeling, which doubtlessly springs from Weart's training, just as he says it does, nor do we take issue with the argument for global warming, which we find fully persuasive. What interests us, rather, is the way probabilities and facts come together—indeed, can be confused with each other—in a good deal of the thinking that argues for global warming. Without agreeing with those who seek to debunk that argument, we do want to address this question: What status have the knowledge claims for global warming? Before addressing this question, however, it is instructive to look at a few elementary matters helping to shape the consensus, at least in broad outline.

The consensus is only recently emergent. As late as the mid–1970s a central issue among climatologists was whether Earth is warming or cooling.[56] There certainly was no compelling case for global warming, and between the 1940s and the 1960s the global trend, in fact, had been lower temperatures.

Cooling does remain a possibility for some even today, but only at the fringe of the scientific community. One major obstacle to the belief in global warming persisting up to the end of the twentieth century was the long-held idea that nature changes ever so slowly over the long course of geological time. Consistent with Darwin's carrying forward the geologists' earlier case for uniformitarianism over catastrophism and his making the argument that "*Natura non facit saltum*," nature makes no leaps, evolutionary thinking until lately consistently fostered the belief that large-scale historical change is glacial in character. Of course, the glaciers themselves have been melting rapidly before our eyes, a fact giving many scientists reason to pause. Then, too, between the 1930s and 1950s historical evidence for rapid climate change arose with the discovery and confirmation of the so-called Younger Dryas. Occurring some 13,000 years ago, this climatological shift reversed a warming pattern and plunged the climate back into ice age conditions virtually overnight.[57] Nature could and did make leaps, which is fully consistent with the theory of punctuated equilibrium that Stephen Jay Gould and Niles Eldredge have more recently promoted in the idea of sudden environmental change giving rise to rapid biological evolution. Some important rethinking about nature's historical record of change was therefore already underway when the consensus took shape. The argument for global warming, nevertheless, needed to make its own case.

Those presenting the case have had somehow to show persuasively, short of proving conclusively, that current warming is something other than a temporary phenomenon — that warming is, in fact, on a trajectory heading towards a projected range of temperature increase, that the trajectory is fueled primarily by human agency, and that unwanted consequences are in the making. Naysayers can buy into belief in warming without purchasing the anthropogenic portion of the argument. Some opponents of the anthropogenic position have argued that variable solar activity can explain warming.[58] Others have simply embedded current warming within a larger historical period of climate change. They believe disappearing glaciers to be merely the closing act of the Holocene epoch, which encompasses the past 13,000 years of melt-off from the last significant Ice Age.[59] Grounding the case for anthropogenic factors, on the other hand, is the "Keeling curve," named for the scientist who in 1960 began measuring and spotlighting the increasing amount of carbon dioxide in the atmosphere. Then, near the end of the twentieth century, a most significant finding emerged, giving important testimony on the issue. Critical to the emergence of a consensus around global warming and its anthropogenic sources was the ice core drilling done at the Vostok station of Antarctica in the 1990s. The Vostok ice record, which exceeded two miles in depth, showed that Earth is almost as warm now as at any point in the past 420,000 years; that a further increase in temperature will place us at the verge of an entirely new climate regime; and, most significant of all for the warming argument, that fluctuations in temperature and fluctuations in atmospheric carbon dioxide historically correlate.[60] Weart tells us this about these findings: "The Vostok results were definite, unexpected, and momentous.... The Vostok core tipped the balance in the greenhouse effect controversy, nailing down an emerging scientific consensus: the gas did indeed play a central role in climate change."[61]

The Vostok findings gave immediate credibility to Jim Hansen's early prediction that a doubling of carbon dioxide would lead to a temperature increase of nearly 4 degrees Centigrade. A scientist at NASA's Goddard Institute for Space Studies, Hansen performed groundbreaking work by bringing issues of global warming into the realm of world modeling. Although 3 degrees later would become more widely accepted, those coming to consensus generally fell in line behind Hansen's work, and climate modeling has become foundational for the argument for global warming. The consensual estimates of twenty-first century warming collect within the 1.5- to 4.5-degree range, and they are all products of global simulation studies.[62] As one group of scientists and science writers tells us, "Scientists who study changes in the earth system cannot follow classical experimental methods because we have only one earth: we do not have a second one without elevated levels of greenhouse gases in its atmosphere to observe for comparison. Instead, our control standard for checking our hypotheses is a computer model that captures much of the complexity of the earth system."[63] Such experimental work and the knowledge claims arising on these experimental grounds rest on no Laplacean bedrock. Their status is something different from that accorded traditional scientific knowledge claims that can be demonstrated. The experimental knowledge arising with planetary simulation reflects directly the real limits climate change imposes upon predictability.

Leaving aside the altogether profound question of what humans may or may not decide to do vis-à-vis global warming, a matter Weart properly judges the most significant uncertainty of all,[64] we need to understand that climate itself is a microcosm of the unpredictable and uncontrollable interactional manifold of nature. The interactional complexity and non–linear causality of our planet's climate condition our knowledge profoundly and make our knowledge claims, whatever their evidentiary base may be, something other than certain. Climate modeling originated, unsurprisingly, in the belief that climate is predictable and controllable, and thus available for military use. This belief belonged to Princetonian mathematician John von Neumann, who helped pioneer the Cold War's early development of the first computer models of a world system. These models established the blueprint for all successive modeling, whatever later refinements might involve.[65]

Modeling divides the globe into three-dimensional cells, entailing latitudinal, longitudinal, and vertical atmospheric chunks that are layered.[66] Data gathered by satellite are critical to the atmospheric dimensions of planet modeling, so the Cold War fed the possibility of climate simulation in more ways than one. Entering data into each cell relies on what is known about the physical, biological, and interactional properties of that particular chunk of reality. Because all such features in reality are far too numerous and complex to ascertain, much less understand, parameters are given mathematical values that constrain the modeled reality. Parameters render the "real system" experimentally manageable by ignoring some matters, covering up our ignorance of many causal interactions, and leveling out reality by averaging particulars within a given analytical cell. "Parameterizing" reduces reality's interactional complexity to the manageable proportions of manipulative design. Testing any given model's applicability to the real world, determining whether or not it is a good design match, is most fundamentally a question of retrodictive testing: when we test-run the model's past, does the design reach a relatively close description of what we now see before us in the real world? If the model matches up in this way, it has strong experimental standing.

The problem with retrodiction in these experimental cases is that the design itself is manipulable in abstraction from the reality it is meant to model. Such testing operates in some ignorance of the Earth's intricate dynamic causalities; the simulation does not really "know" what actually brought about the present state of what is being modeled. For example, we do not know the oceans' capacity to sequester carbon dioxide, nor do we know how changes in cloud cover might influence the troposphere's temperature, two matters of obviously major significance to global warming.[67] What counts in retrodictive testing is having the design closely approximate the reality being modeled. If the match with reality is wanting, parameters can be adjusted or changed to suit the situation. Making the model "fit" in this way means *calibrating* it, which amounts to a subjective determination of what may bring the model into better correspondence with what is being described. Differing calibrations may achieve the needed adjustment or relative fit while having different implications for how the model got there, so to speak. Weart comments on the development of such modeling, as follows: "While the best GCMs [global climate models, global circulation models] could now reproduce the present climate, that was far from guaranteeing they could reliably predict a quite different future climate. The models got the present climate right only because they had been laboriously 'tuned' to match it, by adjusting a variety of arbitrary parameters."[68] Retrodiction generally offers something far less than firm Laplacean grounds for making predictions.

Weather forecasting is notoriously risky, with the likelihood of success diminishing rapidly with the temporal trajectory of the forecast. Forecasts quickly become quite worthless, in fact, after a few days out. The modeling explanation of this fact springs directly from the fundamental insight giving rise to chaos theory some decades ago.[69] Meteorologist Edward Lorenz was simulating weather when he discovered, entirely by accident, that an infinitesimal alteration in the numerical expression of initial conditions of a modeled weather system would significantly change the course of the simulation. He learned, moreover, that this initially minute difference in the initial conditions of two modeled systems would radically spread the degree of divergence between the two, such that all similarity would disappear as the simulations projected further into the future, leading to the so-called butterfly effect. In weather forecasting two weeks is the absolute limit beyond which all prediction in principle succumbs to chaos.[70] One might, on the face of it, assume that climate forecasting only magnifies this problem of prediction, given the overwhelming impossibility of ever accurately specifying the initial conditions of an entire planetary system. But climate modeling numerically averages weather over periods of time, freeing itself for the most part from the short-term sensitive dependency of weather forecasting on initial conditions. Climate forecasting does, however, have problems of its own that inhere in the very nature of climate change.

In climate modeling, difficulties concern, among other things, matters of complexity eluding complete mathematical manipulation and opaque to forecasting. Complexity theory speaks to some of this modeling difficulty in the idea of complex systems. Complex systems entail emergent phenomena that, in Bertalanffy's terminology, are non–summative; phenomena that can emerge spontaneously and whose causality can neither be predicted from knowledge of the system's properties nor be considered a mere aggregate of those properties' interactions. Clouds may be considered an example of a complex system, and it is widely admitted that our understanding of clouds, of their relationship to aerosols, and of their general role in climate change is inherently problematic.[71] A further problem in modeling complex dynamical systems lies in the practically infinite number of characteristics defining the local reality of any analytical cell making up the model. Attempts to refine the resolution of the model by increasing the detail of local characteristics has inherent limits that do venture upon chaos. The mathematician David Orrell writes of this problem in his *The Future of Everything: The Science of Prediction*: "Models become more and more refined, but as more detail is added, the degree of uncertainty in the parameters explodes. Because the system cannot be reduced to first principles, we must make subjective choices about parameter values and the structure of the equations. The models often get better at fitting what has happened in the past, but they don't get much better at making predictions."[72] This particular shortcoming reproduces the asymmetry between retrodiction and prediction. Modelers attempt to sidestep that asymmetry as best they can by processing a scenario along various design tracks offered by different climate models. A robust conclusion arrives when different models converge on one predicted scenario. If they all seem to be saying the same thing, greater certainty can be felt in the likelihood accorded that eventuality.[73] But each model in its own undeterminable way suffers model error.

Model error loosely refers to some discrepancy between the model and the real system being modeled. As a model runs into the future, model errors can grow in a cumulative, dynamic manner.[74] Hence, the longer the projection, the more ramified the future mismatch between design and reality. None of this is news to those doing the modeling—nor should it be to us, given what we discussed in connection with feedback control. The modelers' business, nevertheless, remains forecasting scenarios: offering glimpses of what the future will likely be like at this or that projected time, assuming this or that about greenhouse gas emissions, this or that about Earth's dynamical systems and their feedbacks, this or that about public policy choices, and so on. Over and against this norm of the modeling business Orrell points out that investigating model error is like looking into what he calls the "shadow side of science." "Instead of saying what we can do, trumpeting our superior intelligence over brute nature, it shows what we cannot do. It demonstrates ignorance instead of knowledge, mystery instead of clarity, darkness instead of light."[75] Yet planet modelers generally do not see things from this angle since it counters the very nature of their enterprise. Orrell later notes how keeping global-warming forecasts within the canonical 1.5- to 4.5-degree range helps to stabilize the image of climate prediction and, as he quotes Steve Raynor, an Oxford scientist, to "'domesticate climate change as a seemingly manageable problem for both science and policy.'"[76]

These interesting judgments highlight the temptation to assume complete objectivity, an assumption to which some planet modelers may fall prey in their work. The distinction between forecasting scenarios and predicting the future offers no real insulation against such temptation. But the direct business of planet modelers making the case for global warming is neither planet management nor certain knowledge. They focus primarily on charting the probable course of global climate under these, those, or the other specified conditions. Given the limitations imposed upon simulation-based knowledge claims, the argument for global warming needs couch its scenarios in probabilistic terms. Since the argument addresses various publics and government agencies, statements of probability find expression in conventional terms: something having a probability between 66 percent and 90 percent is *likely*, whereas anything having a statistical probability between 90 percent and 99 percent is *very likely*.[77] Reports on climate change usually employ these everyday conversational terms.

The most significant and widely discussed reports on climate change are those published by the Intergovernmental Panel on Climate Change (IPCC), which in 1988 was born of collaborative efforts by the World Meteorological Organization and the United Nations Environment Program. The IPCC employs over two thousand scientists whose work is to determine everything known about climate change and to assess the future by means of planetary modeling. Based on that work, the IPCC has made four reports: in 1990, 1995,

2001, and, most recently, in 2007. The case made for global warming in 2001 bore the judgment *likely*; six years later the reported case was *very likely*.[78] Modelers had become less uncertain about their argument in general. At the same time, however, the 2007 report expressed more uncertainty than before about various particulars. The 2001 report had spoken of the increasing severity and frequency of storms as evidence of global warming, but the 2007 report openly backed off that assessment. In fact, it stated that there is no clear trend evidenced in the frequency of cyclonic storms.[79] More generally, a number of scientists were becoming concerned that various aspects of climate change actually had a greater likelihood than that assigned to them in the IPCC's reports. Hansen himself saw conservatism in the 2007 report, especially projections regarding the likelihood of polar melt.[80] Political considerations can and do enter into what the IPCC puts into its reports, but mistaken projections cannot be reduced simply to the conservative hedging of bets. What should be anything but surprising, in fact, is nature's refusal to follow the modelers' simulation trajectories. In speaking of events' being likely or very likely, IPCC reports have witnessed nature's indifference to their projections from the very beginning: indeed, the actual rise in sea levels has exceeded the IPCC's 1990 projections by 50 to 100 percent.[81] Small wonder, then, that those reports speak in terms of likelihoods of this or that degree of uncertainty.

Arguing what is likely or very likely returns us to the question of the status of knowledge claims about global warming. Earlier the physicist and science historian Spencer Weart identified the very likely probability of global warming as a fact. Apart from a probability's being a mathematical fact, such phrasing confuses, especially in view of the conflation of fact and belief found not uncommonly in the arguments for global warming. Naomi Oreskes has made an excellent case for global warming in a very fine essay, "The Scientific Consensus on Climate Change: How Do We Know We're Not Wrong?" which, nevertheless, exhibits such a conflation of belief and fact. She writes this: "[I]t is no more a 'belief' to say that earth is heating up than to say that continents move, that DNA carries hereditary information, and that HIV causes AIDS."[82] Yet a mere two sentences later Oreskes goes on to tell us the following: "Might the consensus on climate change be wrong? Yes, it could be.... This possibility can't be denied."[83] Oreskes' understanding anchors itself both in certainty and in uncertainty. In exactly this same vein, let us note the position taken by Joseph F. C. Dimento and Pamela Doughman in their essay on what global warming will mean for our children and grandchildren. "The future of climate change is a phenomenon that is not knowable with certainty, but current and probable trends are risky to ignore."[84] And then, a mere three sentences later, they tell us this: "Denying that global warming is real is simply a refusal to look at the evidence."[85] Looking closely at the thinking of those making the case for global warming can lead to head-scratching. Is global warming really a fact, or is global warming perhaps something other than a fact? Is global warming a reality, or is it simply an argument about what is very likely a reality? As persuasive as we find them, those making the case for global warming often place themselves on both sides of these questions.

The scientific consensus is that global warming is altogether real, yet the scientific knowledge we possess does not permit the scientific community the same claim of certainty commanded by matters having factual status. Of course, the case for global warming consists in many facts having great significance. Various climate records offer foundational support for the global-warming argument, which holds to anthropogenic causes, and along with all such globally measurable data there are the factual events we see and hear about either directly or through the news. Significant numbers of species having millennia-long niches are in migration, abandoning local habitats in search of new niches in different climes. Species enjoying less adaptability are dwindling towards extinction, with many already lost. Hibernation and procreation patterns are changing. Spring is coming weeks earlier than before in different locales. Sea levels are rising. Then, of course, we witness the glaciers' continuing melt, Greenland's rapidly dwindling ice sheet being the most remarkable. All kinds of facts, to be sure. But by themselves these facts are not conclusive. They do not allow us to make certain knowledge claims respecting the future. And yet there seems to be certainty in the minds of so many making the case for global warming, as well as in the minds of those belonging to the consensus, no doubt. What underlies such certainty?

Some would maintain that the facts really do speak for themselves and that a probability amounting to "very likely" psychologically warrants feeling certain. We can allow the psychology of probability without assuming that facts speak entirely for themselves. Counter to this idea that, left to themselves, the facts make the case, we want to suggest that the certainty evident in so many arguments for the reality of global warming springs from two sources that together relate to the facts without reducing to them. The case for global warming entails elements of belief that lean heavily on the facts but draw their sustenance from common sense. It is on his training as a physicist and science historian that Spencer Weart bases his "feeling" for where scientific claims can be assumed reliable, but for most who believe in the reality of global warming, it is common sense that roots their belief. The case made for warming draws our attention to factual evidence, and rightly so, and in so doing, its appeal to common sense is made without being seen. This appeal is ever so subtle, yet ever so significant, as becomes clear in the fine piece by Naomi Oreskes quoted from a page or two earlier. Forgetting that she has just stoutly declared that global warming is not a belief, Oreskes tells us why we should believe that global warming is real and underway, as follows:

> Should we believe that the global increase in atmospheric carbon dioxide has had a negligible effect even though basic physics indicates otherwise? Should we believe that the correlation between increased CO_2 and increased temperature is just a weird

> coincidence? If there were no theoretical reason to relate them and if [numerous scientists earlier] had not predicted that all this would happen, then one might well conclude that rising CO_2 and rising temperature were merely coincidental. But we have every reason to believe that there is a causal connection and no good reason to believe that it is a coincidence.... We have changed the chemistry of our atmosphere, causing sea level to rise, ice to melt, and climate to change. There is no reason to think otherwise.[86]

Here we see belief clothed in conviction, and the terms of the argument are quite plainly those of common sense. Were we to improve on Oreskes' statement, it would be only to add the modifier "good" to the noun "reason" in the final sentence quoted: there is no *good* reason to think otherwise. That addition speaks even more directly to common sense. The point, in any case, is that while our knowledge falls short of objective certainty, common sense compels belief. And the term "certainty" really should be abandoned here where we are actually speaking of *conviction*, which is the altogether appropriate term in this context for the connection between common sense and belief. Certainty is a term pertaining to issues of knowledge, as we have maintained from the beginning pages of this study. "Likely" or "very likely" respects questions of knowledge, and, in that domain, those arguing global warming necessarily admit uncertainty. Our discussion here, however, really concerns the conviction of those who admit uncertainty in their knowledge claims while, at the same time, maintaining that global warming is a fact; it is real. Conviction is the telling term precisely because it allows for uncertainty in the realm of knowledge without forcing abandonment of belief.

We do not at all question the belief of those arguing for global warming; we share it. We also find it entirely appropriate to appeal to common sense in cases such as this where we lack the objective certainty to prove what we fully believe to be the case. Clearly there are those scientists opposing the consensus who, apart from any question of their arguments possibly being employed for nonscientific interests, do not find the case for global warming at all compelling from the standpoint of common sense. They also may be subject to poor or unfair treatment as a consequence of their critical posture. So, too, there are, on the other hand, those scientists holding to the consensus who do so either from pressured self-interest or simply out of conformity. The real world is messy in such ways, but none of this sullies the larger argument being made here. Yet we are not quite done with the facts or through with the discussion of common sense. There is one more point to make, the second in our argument here, which also relates facts to belief without altogether reducing belief to factuality. And this point, like the one just argued, does not stand out or announce itself in any way. Just as unstated and subtle is the quiet reliance the argument for global warming has upon its narrative elements, which encase that argument in a story that is still very much underway.

Looking closely at the relationship between fact and belief in the argument for global warming, we can discern a mixture of the two quite different modes of scientific understanding we have discussed in this study — the positivistic and the narrative. These traditions do not shake hands methodologically. The overarching tradition of positivism, which wants to govern all of the natural sciences, demands a certainty that the experimental control of nature alone can guarantee in its command and prediction. The narrative tradition, which, as a minority tradition, has had to defend its scientific credentials on various professional fronts during the last half-century, acknowledges natural causalities that elude our manipulative grasp and admits uncertainty in its knowledge of a future that cannot be fully predicted. The argument for global warming in some sense falls between these two traditions; or, better, it conflates them. On the one hand, it offers factual evidence, much of it obtained through highly objective research, and it partakes in global simulations. On the other hand, the argument entails the story of global warming that draws all such evidence together within a meaningful framework that remains open-ended. Those arguing the case understandably present their evidence in the best scientific manner, concluding with probabilities that appeal to our need for objectivity. The narrative dimension of the argument works differently, however, interrelating the factual evidence along with forecasts and shaping it all as a persuasively meaningful story that, like all good storytelling, pitches its appeal to common sense.

Scientists arguing for global warming speak from the positivist tradition in which they have been acculturated while, at the same time, they are having to refer to natural developments which escape the methodological confines of that tradition and call upon the narrative mode of understanding. Scientists are telling a story that is still underway. Without any presently given conclusion to the story they are telling, they speak of our being in the midst of an unfolding narrative. Nature supplies the material of the narrative in the way of factual reality. Glaciers are not melting in the story alone; they are melting in fact. Oceans are not rising in the story alone; they are rising in fact. The Vostok findings are not fictionally established for narrative purposes; they are recordings of nature in actual fact. Our conviction draws upon all such evidence, to be sure, but it actually has something to do also with the story itself, for it is the narrative mode of understanding that brings together these facts in a sharply defined coherent explanation that has compelling force. In short, the story as a whole lays some claim to our judgment. There is something compelling in the way the evidence all fits together, not as a snapshot or picture-puzzle, but as a story that joins past, present, and future in cogently coherent and forcefully persuasive meaning.

The point here stands out most strikingly against the negative. Opponents of global warming often have attacked the argument on Laplacean grounds, knowing that demands for decisive objectivity and predictive certainty cannot be met. What is most telling about the naysayers' position, however, is that they lack one. If we consider the naysayers as a population of scientists, it is solely their opposition to the argument for global warming that brings them together; there is no further unity because they disagree among themselves on fundamental matters requiring explanation. Each naysayer takes a position of some kind, but be it an argument for solar activity, for a continuing Holocene, or whatever, it is not a position finding any widespread agreement.[87] Critical it may be; persuasive it is not. Not one among the naysayers has been able to effect any widespread agreement in the scientific community, and this fact alone carries considerable force. Over and against such disunity stands a consensus among the majority of the scientific community, a majority for whom the argument for global warming carries its own force, leading most to conviction. Consensual validation guarantees no truth, of course, as the history of science well illustrates. But in the absence of demonstrable proof, it is the story of global warming that alone widely satisfies the need for a coherently convincing and convincingly coherent scientific explanation of all the facts, taken together and considered as evidence. This too is fundamentally a matter of common sense.

Scientists will go on developing the story, of course, and well they should, just as global simulations will continue playing a necessarily central role in our knowledge and expectations. Objectivity and storytelling will still maintain their strange coexistence, but whether or not forecasting will extend credibility for the larger narrative will turn, in part, on actual events and developments out there in the real world, so to speak. But, then again, global warming itself may never provide any clear-cut development or decisive occurrence clinching the case. The tendency to look for clear empirical confirmation of the argument is understandable, of course. Scientists more than once have deemed severe weather events like drought in Sudan and Hurricane Katrina in New Orleans clear evidence of global warming, and the 2001 IPCC Report fell prey to this same temptation. Yet weather variability alone also can explain such singular occurrences, and when the critics of global warming point out such possibilities in order to besmirch the argument for global warming, no good service is done to anyone's understanding.

New York Times science journalist Andrew Revkin notes how critical mistakes of this kind more likely than not create public confusion, which is especially a risk in the United States, where considerable public ignorance persists as to what scientists do or do not believe when they speak of global warming. As Revkin writes about warming more generally, "After covering climate for over twenty years, my sense is that there will be no single new finding that will generate headlines that will galvanize public action and political pressure. Even extreme climate anomalies, such as a decade-long super drought in the West, could never be shown to be definitively caused by human-driven warming."[88] Revkin's judgment follows not only from his understanding of the non–evidentiary status of singular weather events, but also from his professional experience with newsmaking. Eventfulness is what grounds newsworthy items, and seemingly discrete environmental events generally cannot disclose developments unfolding temporally, even if the story connecting the events is the most significant news story of all. The newscasting of events is like photography; it takes snapshots. The story of global warming is a film; it takes time. Frames of that film provide snapshots, but the snapshots shown by newscasters do not in and of themselves disclose the film's ongoing story. Scientists and environmental groups often strive to offer scientific snapshots as evidence of the story they are telling — which recapitulates the mixture of Laplacean vision with narrative understanding in the argument for global warming. Hence, we come round again to the problem of uncertainty, only now it appears from the standpoint of newsmaking. And Revkin's advice to journalists on the science and environment beat is absolutely spot on and well worth hearing on this score.

> Journalists dealing with global warming and similar issues would do well to focus on the points of deep consensus, generate stories containing voices that illuminate instead of confuse, convey the complex without putting readers (or editors) to sleep, and cast science in its role as a signpost pointing toward possible futures, not as a font of crystalline answers.... By getting a better feel for the breakthrough and setback rhythms of research, a reporter is less likely to forget that on any particular day the state of knowledge about endocrine disruptors, PCB's, or climate is temporary. Readers will gain the resolve to act in the face of uncertainty once they absorb that some uncertainty is the norm, not a temporary state that will give way to magical clarity sometime in the future.[89]

Revkin's advice to journalists exhibits wise judgment far beyond the normal confines of journalism or the single issue of warming. Indeed, his judgment really speaks to the heart of a sensibility befitting our living beyond any Promethean illusion in an age of uncertainty. Where Revkin speaks of newsmakers, we should include the myriad men and women who address nature as scientists and who tell us about what they do, what they know, and what they understand about our relationship to nature. And where Revkin speaks of readers, we must think of ourselves as needing an awareness conforming to his advice. Learning to live with uncertainty in such a fashion, if we are to learn at all, will likely take time, and time itself is at a premium for most of those who want to effect meaningful change. While the story of global warming lacks the express causality and eventful visibility of newsworthy affairs like Beijing air pollution or the effects of Chernobyl, many scientists and environmentalists want to activate public interest and pressure public agencies on all levels immediately, before warming reaches certain tipping points that could irrevocably make the story's material far worse in coming chapters. Unable to produce "crystalline answers" or provide "magical clarity," what are they to do? Lacking Laplacean vision and the ability to control natural processes, these scientists increasingly are finding themselves in the role of storytellers, which is a role for which the reigning positivistic tradition of modern science has hardly equipped them. Yet storytelling is very much an appropriate mode of scientific understanding in this land of sometimes and perhaps, where science no longer commands the realm of appearance and uncertainty is widespread across the land.

Conclusion: Rethinking Our Being with Nature

Global warming is but one of possibly myriad stories underway. Revkin in his essay refers as well to endocrine disrupters and PCBs, and, of course, there are emergent diseases and a whole host of other materials and processes that may be interactionally shaping significant narratives even as we speak. In any event we are caught up in stories: some old, some new; some weighing light in the way of factual material and persuasive coherence, others weighing heavy. Whatever the case, neither our culture nor the scientific community is well schooled or long experienced in approaching practical questions of our relationship to nature on grounds lacking full objectivity. The very authority of modern science crystallized three centuries ago with practically certifiable knowledge claims, and knowledge falling short of full objectivity still remains suspect across large segments of the scientific community. What constitutes a scientific consensus on global warming is therefore truly an exceptional moment. In this singular case where the stakes are so very high, scientists have managed to place one foot outside the long shadow of Newtonian science; they have taken a position on the question of global warming without giving in to the demands of positivism. One would hope that this exceptional moment is a portent of an emerging sensibility. A more sober judgment, however, may be that this moment is simply exceptional.

While judgment on the question of global warming has taken scientists one step outside the reigning positivist tradition, that tradition remains generally very much intact in calls for instrumental answers not only to warming, but to numerous other problems of climate change as well. Instrumental knowledge certainly must play a critical role in how we respond to the worsening problems of climate change. Taken together, however, those problems in themselves call for more than an instrumental "fix" in our relationship to nature, such a fix being ultimately no solution at all to the dangers inhering in our illusion of mastery. Moreover, a larger problem may well be that, as our problematic interactions with the natural life-world increase and worsen, proposals to remake nature and seize control of global interactions could have widening appeal and deepening thrust in a culture still bound to that illusion. Indeed, the very power of this illusion — the temptation of power itself—could increase in the face of the crisis of the natural life-world. In the conclusion that follows we will turn our understanding toward the future to consider briefly how we might think differently about nature in a world that has come to its senses. Pushing the point of this study beyond the past and present, we want to consider what sensible understanding might look like once thinking about nature no longer has any modern project of mastery in mind.

The few brief reflections to follow keep faith with what has gone before, not only in particular respects but also in their general character: they do not pretend to provide any answers to the crisis of the natural life-world. Thinking sensibly does not mean solving problems, at least not fundamentally, but recognizing from the ground up that our being in the life-world means living in a land of sometimes and perhaps without any instrumental compass to guide us safely or surely. Learning to live with this reality—this uncertainty—is what sensibility should mean in its most elementary and essential character. What follows are simply a few reflections on what such sensibility might entail, nothing more, nothing less.

Towards a Sober Appreciation of Nature

In her book *The Human Condition*, which she published in 1958, the political thinker Hannah Arendt at one point reflected on the impact the natural sciences have had upon our relationship to nature in the modern age. She wrote of how, through the introduction of experimentation — through the impressments of "man-thought conditions" on natural processes— the moderns recently have managed to force nature and its energies into processes and patterns that would never have arisen without human artifice.[1] As tool-bearing animals human beings have always been able to act instrumentally towards nature, but the instrumentalization of human action arriving with modern science so widened and deepened our penetration of nature that now we are "making" nature and "creating" unprecedented natural processes. Arendt's point, however, was not to publicize the burgeoning mid–twentieth-century achievements of science or marvels of technology, matters already quite evident and much ballyhooed at the time. Nor was she interested in making a case for the new epochal sensibility Paul Crutzen later articulated in the idea of the Anthropocene. Rather, she called attention to the character of human action and to the fact that, even when instrumentalized, *action* will have its own way.

Action is for Arendt primarily a political concept. At the same time, it is more primordially a kind of existential category; existential in the loose sense, that is, of helping to delimit and define the "human condition" through capacities identifying what it is to be human. Whereas human *behavior* refers to things prescribed or routine, activities essentially predictable, human *action* refers to the undertaking or initiating of new things. It refers to beginnings.[2] The capacity to act corresponds to human natality: to act is to bring something unprecedented into the world, to enact or initiate something uniquely novel. Action also initiates a chain of events consisting of others' re-actions, one following another in a processual sequence having neither script nor blueprint to channel its course. And it is this indeterminate character of action that counts most for Arendt even for instrumentalized action, which in its nature has nothing properly to do with things political.

> The very fact that natural sciences have become exclusively sciences of process and, in their last stage, sciences of potentially irreversible, irremediable "processes of no return" is a clear indication that, whatever the brain power necessary to start them, the actual underlying human capacity which alone could bring about this development is no "theoretical" capacity, neither contemplation nor reason, but the human ability to act—to start new unprecedented processes whose outcome remains uncertain and unpredictable whether they are let loose in the human or the natural realm.[3]

Pursuing matters lying in other directions, Arendt's thinking left science and nature behind. The theoretical angle of her vision was far too acute, in any case, to enjoy widespread appeal. Arendt's prescience here is remarkably illuminating nonetheless, coming as it did before anyone had publicly spotlighted the ecological material for which her point might have provided some theoretical coverage. Her words speak to our time: not so much to the Anthropocene as to our living in an age of uncertainty. "In this aspect of action ... processes are started whose outcome is unpredictable, so that uncertainty rather than frailty becomes the decisive character of human affairs."[4]

It is entirely possible that uncertainty will become—indeed, that it is already becoming, albeit ever so slowly and imperceptibly—the central and defining feature of our consciousness of our relationship to nature. The biological and ecological material we addressed in the last few chapters is but a minute sampling of the interactional dynamics of the natural life-world. Yet even those few cases do illustrate quite well how our interactions with nature continually effect unpredictable change both behind our backs and right before our very eyes. As Colborn and her associates warned in connection with our regular ingestion of chemical soup, there is no telling when or where the next nasty surprise will emerge, and the only certainty is that we will indeed be blindsided again and again. With similar blindsiding in mind our medical antennae are ever alert, but SARS and HIV evidence how no measure of medical intelligence and research can foresee, much less forestall, microbial surprise attacks. Even in the case of global warming, where at least we believe we have some general understanding of what is occurring, unexpected twists are most likely to happen. As time passes, ongoing environmental study and continuing climate reports may help paint the picture of what lies ahead with more confidence; yet, at the same time, a great many of those brushstrokes will generally be broad and not necessarily revealing either regionally or locally, where uncertainties may well heighten and proliferate in particular cases. Moreover, something altogether unexpected could occur and alter the very contours of expectation.

On the ground floor of rapid environmental alteration and on every level of organic change, as species disappear, ecological niches alter, and interactional patterns of the life-world assume fluid and unprecedented configurations, there is no telling what contingencies and novelties will increasingly shake our normal confidence that tomorrow will look like today. At some point such shaken confidence could, in turn, help us realize along with Hannah Arendt that uncertainty is the decisive characteristic in human affairs, or at least in those affairs connecting us to the natural life-world. Understanding that tomorrow probably will not look like today will make yesterday and today questionable guides for tomorrow, and this eventuality ought to at least make belief in progress an extremely difficult compass to follow in the coming decades. Yet any such assumption may well be far too facile given the illusion of mastery still prevailing in our culture. Indeed, the two leading programs of instrumental adaptation promote one kind of mastery or another as an answer to the mess we are making of this planet. But the illusion of mastery hardly restricts itself to grand programs of self-conscious control, wherein our species knowingly takes charge of its own evolution and/or that of the planet's future. The conventional illusion of our domination of nature also lives on generally in the more prosaic and diffuse faith most of us tacitly have in the instrumental ingenuity lying behind our science and technology. By and large, we do trust in our instrumental capacity to handle virtually any challenge nature may throw at us. It is, after all, precisely such instrumental ingenuity that built the very material world we inhabit, and did so pretty much in piecemeal fashion without any Baconian House of Solomon behind it. Nature had little to say against any of that, and certainly any obstacles it threw at us proved temporary and anything but insurmountable. Given the faith in mastery this stunning world of human artifice inspires, whether we adopt conquest by design or something less programmatic, we seem to have in science and technology a constant instrumental bridge between today and tomorrow. Instrumental knowledge can provide solutions one way or another: it can fix the problem(s). But of what substance is this instrumental bridge made outside of illusion? Or, turning this question's thrust in the positive direction we wish to follow here, how should we think about nature without the modern assumption of mastery in mind?

Belief in domination has generally so conditioned our perception that our reflexive response to almost any problem with nature typically is "fix it." The assumption governing our perception and response is that we can command nature if we put our mind to it. Hence, the very appearance of a problem with nature is framed so as to induce the automatic reaction to solve it. Problems entailing uncertainty essentially are subject to the same conditioning, or framing: assume control of the problem, fix the problem, eliminate or manage uncertainty, and thereby eliminate or manage risk. The issue for us here is not whether or not particular problems are safely fixable; the more that are, the better. We surely are not arguing against practical solutions, as if we should shun technologies properly designed and sensibly employed to handle treatable problems. Solar technologies of all sorts must be developed and refined, and a vaccine for HIV or the Ebola virus certainly would be something wonderful to celebrate. The critical point here, rather, is the illusion giving rise to a virtually autonomic response to problems with nature. Seeing through that illusion calls for reconditioning our response. We need a new reflex consistent with the absence of control and the nature of uncertainty. That reflex is prudence.

Prudential concerns are anything but novel in discussions about environmental problems. Immediate with growing attention to environmental disorders in the 1970s came the requirement of preparing environmental impact assessments for newly proposed intrusions into nature. Apart from questioning just how seriously or meaningfully such requirements were undertaken or enforced, the idea behind those assessments spoke to prudence. Not surprisingly, given the bent of our larger instrumental sensibility, the idea of prudence automatically inclined towards objectivity. Prudence was operationalized in calculi of probabilities respecting possible risks, costs and benefits, and the like. Not that such calculi were or are necessarily clear in themselves or easy to apply. Indeed, those promoting the "precautionary principle" along such lines often seem conspicuously conscious of their merely gesturing in the direction of prudence while suffering incalculable problems in operationalizing what prudence might mean in any particular case.[5] Those problems are real, and where costs/benefits and risks are clearly determinable, those efforts are not to be disparaged. But from our point of view in this study, the ironic risk of all such endeavors—it was less a risk than a foregone conclusion—is that prudence would become one with management: risk management.

Risk management has become such an industry that now, in this "meta" world of ours where everything becomes elevated by mirror reflection at a higher level of abstraction, we have serious professional concern for the management of risk management, which has to do with how controversies over questions of hazard (risk management) are themselves to be managed (risk-issue management).[6] At bottom, of course, the fact is that we were bound to lose the essence of prudential wisdom, Aristotle's *phronesis*, once instrumental reason laid claim to its meaning. The point is that prudence itself cannot ever be a fix; it is never a solution. Yet risk management commits prudence to just such understanding, at least in principle if not in fact. In calling for a prudential reflex, what we have in mind is nothing calculated or even reflective; hence, the term *reflex*. Prudence must in some sense define the character of our sensibility such that a real concern for danger first comes to mind; whether or not there is any good reason in a particular case to assume such concern, it simply is there automatically. Unmediated by thought, this caution is simply the *given* in our response. And in this same regard, *danger* or some equivalent word would be much more apt than "risk" in our thinking about a prudential reflex. It is more expressive and telling of what we have in mind. This point has been made from a somewhat different angle by Langdon Winner, one of the most thoughtful students of our relationship to our technologies. He insightfully observes that the very term *risk* favors those interested in going forward with the project or intrusion; the word itself conveys the point of possible gains and benefits, and, moreover, we live in a culture that embraces risk-taking as one of the warrior virtues.[7] Winner's insight certainly rings true in the arguments of those genetic engineers wanting to connect their version of instrumental adaptation to what they see as our species' trait of risk-taking, that trait being in their view responsible for lifting us out of the African savannah and for developing civilization itself.

Danger, in any case, befits the idea of a prudential reflex far better than risk. It holds hands with the fundamental idea that prudence is no solution, no fix; that it is, rather, essentially a posture towards the world that brings with it a reflexive response putting our senses on defensive alert. Prudence does not, however, identify what normative regard for nature we should have. If we are not to see nature simply as material for human manipulation and use, how should we behold and evaluate nature? We know full well the master's dismissive attitude, having discussed it in so many different contexts already. However, the objectification and manipulation of nature is not something we should do without, any more than we would call for an end to experimental research on mortal diseases. Nothing in this study underwrites Luddite thought or denies instrumental knowledge its due beyond the theme of domination. Nor is there in this work any romantic appeal to a simpler time: there should be no illusion whatsoever concerning the killer diseases that used to stalk this Earth. Yet, that said, the Baconian reduction of nature to slave status clearly will not do. Nature simply will not have it that way, nor can our attitude make it so by any instrumental decree. Hence, the question is this: living without belief in mastery or the presumption that we somehow can govern the future, either that of our species or that of this planet, what normative thought or evaluative regard for nature is most appropriate to our interactional dealings with the natural world in which we are enmeshed and towards which we should assume a prudentially reflexive posture?

Offerings of a new understanding of nature abound in the ecological literature that is critical of our abuse of nature and concerned with remediation of one kind or another. At one extreme are deep ecologists who prescriptively refer consciousness to ontological and epistemological frameworks in an effort to restructure our perceptions and understanding. Among deep ecologists are some holding the perspective of the Jains, who grant all living things natural equality within a living community that outlaws damage to life beyond the absolutely barest minimum needed to survive.[8] Coming at planetary crisis from an altogether different angle are a good many critical ecologists who follow James Lovelock's Gaia hypothesis in some fashion. This hypothesis places our interactions with nature within the larger context of the superorganism that is the living Earth, and many holding this perspective call for a monumental reduction of the human population along with an adjustment of human activities to suit the homeostatic functioning of the entire organism.[9] Not all deep ecologists agree among themselves, nor are the promoters of Gaia likeminded on everything. So too, of course, myriad other perspectives exist on the change of consciousness we need to effect for ourselves if we and the biosphere are somehow to survive this mess we are making of Earth. Cataloguing such various perspectives on how we need to see nature differently could take us in many directions, leading us nowhere in particular. In the absence of a common denominator, however, we can at least note one idea that echoes generally throughout much of the literature calling for a change in our consciousness. Most generally, whatever their particular perspectives may be, large numbers of those proposing a new regard for nature speak of our needing to leave behind the outworn machine metaphor of mechanistic understanding, which enabled us to mistreat nature without any second thought whatsoever, and of our needing to think of nature and our relationship to it in biological terms that will allow us to work with nature collaboratively. We need, in short, to assume some organic understanding of our relationship to nature and work in partnership with her.

This understanding, whatever its particulars, is well meant towards nature, and this biological-partnership model can translate into sensible policy proposals at various levels. When considered in the round, however, there are problems with the notion of biological partnership. For one thing, too much is left up in the air. It is not at all clear what partnership with nature should mean for our encounters with nature beneath the abstraction of the idea itself. During the winter season a farmer may reflect on the give and take between nature's ways and his own, but in what meaningful sense would the farmer think about some partnership with nature? If nature delivers horrendous harvests three or four years running and puts the farmer in bankruptcy, the idea of partnership surely rings hollow. Floods, storms, the ruinous tornado—these are not the makings of partnership to those who are closest to nature when they work the land. Nor does the idea of a collaborative, organic relationship speak to the reality of the encounter with nature a medical researcher has while experimenting with some deadly microbe such as the HIV virus. Pitting her hopes and ingenuity against what may seem an implacable foe, that medical experimenter encounters no partner in any sense. We must realize that in the profoundest sense, any partnership in life is at once a partnership in death. Such is nature's way. Stephen Jay Gould puts the point best very plainly where he tells us that "Nature does not exist for us, had no idea we were coming, and doesn't give a damn about us."[10]

Another problem in this call for biological appreciation and partnership is the risk it runs towards confusion, if not open complicity, with proposals to assume some sort of governance over the biosphere or some large portions of it. A possible alliance of that sort surely should come as no surprise since the idea of instrumental adaptation quite openly allows, if not invites, the wedding of biological sensibility and collaborative partnership with belief in mastery. That is a partnership all right, but not necessarily the one we are meant to suppose when we hear talk of our needing to assume some collaborative responsibility towards this planet and the living things on it. Having addressed such planetary ambitions already, let us here simply note that the initial installment of Myers' *The New Atlas of Planet Management* was entitled *Gaia: An Atlas of Planet Management.* Lovelock himself gives no quarter at all to such thinking, whether it pertains to his notion of Gaia or not. As he wrote in a recent work, "The idea that humans are yet intelligent enough to serve as stewards of the Earth is among the most hubristic ever."[11] Gould expressly aimed his point above at the same presumptuous call for stewardship, his own standpoint respecting the temporal scale of geological time and humankind's momentary presence on this Earth. Belief in our needing some biological partnership with nature blends all too nicely with ideas of stewardship, which readily insinuate domination of one sort or another into the very thinking intending to supplant the old mechanistic exploitation of nature. Mastery can find accommodation in this presumption of partnership only because such thinking flies too abstractly above our relationship to nature.

A proper regard for nature actually is not that difficult to identify once we stand outside any belief in domination. It arises almost at once, in fact, when nature is recognized on its own terms, as it were. Let us reconsider the experience of the farmer and the medical researcher imagined above: he who struggles against floods and storms, and she who battles experimentally on the HIV front. They both work with their hands, with instruments as different as the tractor and the microscope. They both intervene in nature by design, be that the spacing and seeding of rows in a field or the design and execution of an experiment. Their languages are different up to a point, as the farmer lives in the vernacular while the medical experimenter operates mathematically. Yet the scientist does return to the vernacular at the end of the day, and if we imagine these two sharing their thoughts on their encounters with nature, it is not hard to understand how each might be completely open to understanding and appreciating the other's experience, at least essentially. This openness, this basic appreciation of each other's experience, might well spring from the respect each one ought to have for nature. Win, lose, or draw, each one's encounter with nature ought to mean respect, be this respect obtained in the face of weather, pests, and ill fortune, or in the laboratory in the face of an incredibly protean microbial reality standing somewhere between the living and the inanimate. Nature commands that respect on its own terms. Yet these terms are evident and this point is clear only in so far as we are mindful of nature's being nature in the first place. Farmers have never had a problem on this grassroots level, traditionally struggling as they have on the land, just to make things work. It has been a different story in modern time, however, where belief in human command has hardly allowed nature any claim to respect.

Quite often we hear that respect must be earned. If that standard be applied to nature, millennia of human testimony suggest that nature earned human respect fairly quickly, probably right from the very start, and managed to maintain it until quite recently. Respect certainly lived wherever fear or awe of nature informed cultural norms. Pre–modern Western peoples and traditional non–Western cultures are univocal in this respect. Nature had everyone's respect but then lost it with the coming of the modern age, at least among those peoples who exercised the sensationally expanding power of instrumental knowledge. That experience was architectonic for Western culture as a whole, and the exercise of instrumental power has intoxicated Western sensibility ever since — all the way from the eighteenth century's *Encyclopedia*, through the Crystal Palace of 1851, to the claims of the genetic engineers and planet managers of today. Belief in our conquest of nature is itself an intoxication, and those drunk on power have little in the way of sobriety when looking at nature. This stupor truly explains why respect has disappeared from our relationship to nature: nature really did not lose our respect; we simply denied nature any right to claim it. What master assumes a slave can demand anything, much less command respect?

The fact of the matter is, of course, nature never really was such a slave, and it has been anything but silently obedient during the past one hundred years. Nature's recalcitrance tore apart the scientific world at its microphysical foundations almost a century ago. Arthropods and microbes today are running resistance rings around our continuing efforts to thwart their nefarious actions. Countless chemicals crisscross in our bodies with known endocrine-disrupting capabilities as well as carcinogenic possibilities, and we possess no clear idea of what life-time will tell us about our own bodies much less about the flesh and blood of our progeny. And so it goes, right on through the unforeseeable consequences of what *very likely* is the anthropogenically based warming of our Earth. Nature speaks loudly in all these cases. What more must it do to earn our respect? But, then again, nature really is not the least interested in our respect. It is above all that; it simply "doesn't give a damn about us."

This nature — not any nature dreamed of by the genetic designers or that of the planet engineers— is the nature that commands respect. The nature of which we speak here, and which by every measure of sober sensibility should command our respect, is the nature that will determine how hospitable or inhospitable this home of ours will be for the generations ahead. What we choose to do now and in the future will have some real bearing on what lies ahead, but we must realize at the same time that the substance and extent of that bearing is itself beyond anything we can determine. Rachel Carson did not know even the half of it when she pointed out almost half a century ago that we have begun a great experiment on ourselves over which we exercise little or no control.

In any event, we must choose to think about institutional changes, policies, and concrete actions with a clear eye towards nature. Those choices lie well outside the scope of this study, but we do hope that this work ultimately has something meaningful to say about the clarity and sensibility we need in thinking about our relationship to nature. David Orrell capsulizes the entire matter quite nicely where he tells us the following:

> [W]e in the industrialized world still tend to see the world in objective terms, as something to be manipulated and controlled, slave to the laws of cause and effect. This is the shadow side of our great inheritance from two millennia of science. Those predictive models of the world not only looked into the future but, in many respects, helped define it. By turning the world into an object we can control, however, we also deny it life.... We will choose to protect nature only if we value it — and not just as an object, but because it is alive. The only way we will respect it is if we understand that we cannot control it.[12]

David Orrell's insight is spot on. Respect for nature and a prudential posture towards nature join hands here in sober sensibility. And we need such sobriety badly, but not for nature's sake. Our need for sobriety speaks to our own need as a species; it is our own well-being that is at stake, along with that of so many other living things.

Our concluding remarks thus far have fallen on one side of the divide typically splitting critical ecological discussions between calls for changes in consciousness and calls for changes in policy and technology. If these be the poles of discourse, the discussion above clearly points towards consciousness rather than policy: prudential sensibility and respect for nature belong more to a concern for normative comprehension than to the policy arena, though these matters ought ultimately to be thought inseparable. Yet we have not completed our concluding reflection, as there remains one more point to consider, a point that in its own special way inclines very much towards policy of a non–instrumental sort. There is one more stop, a very important stop, to make along the path of our rethinking our being with nature.

Redirecting the Project Mastery

A full two decades ago Bill McKibben issued a very timely public alert about global warming in his important work, *The End of Nature*. His concern in that book reached well beyond raising an alarm on rising temperatures. The "end of nature" means the collapse of the distinction between the natural and the artificial. Atmosphere, weather, climate, earth, sea, the chemical constituents of life itself: human action has wittingly or unwittingly altered all of them, and nature no longer has any preserve providing it sanctuary from human interference or influence. McKibben's work is, among other things, an open lament for nature's loss of its independence from humankind. Behind this lament lies McKibben's deeply American appreciation of nature as an unspoiled wilderness. It is the wilderness of John Muir. It is the natural beauty idealized for the imagination by the likes of the Audubon Society's photographs. But it is much more the raw, rugged wilderness that until recently was so independently alive in places like the Adirondack Mountains of upper New York State where McKibben has lived and where this author lived for four years some time ago. In such places the deleterious effects of human action have long been all too evident. We have corrupted the wilderness, consigning much of its meaning to human memory. Indeed, for McKibben, nature's loss of independence means the loss of nature itself: "Nature's independence *is* its meaning; without it there is nothing but us."[13]

McKibben's understanding here suggests the pollution parallel. The collapsing of the distinction between the artificial and the natural is like the mixing of the impure with the pure, and it takes very little of the impure, sometimes only a single drop, to corrupt the pure in its entirety. Sometimes impurities are distillable and can be eliminated, but not in this global case. McKibben's threnody echoes in a future empty of nature. But this pollution parallel poses problems for discussing what, in fact, remains an interactional reality in which *humankind* and *nature* need remain separable. The problem is plain in McKibben's 2005 "Introduction" to *The End of Nature*, where he writes: "Hurricanes and thunderstorms and tornadoes become not acts of God but acts of man. That was what I meant by the 'end of nature.'"[14] It is unclear exactly where such meaning thrusts. In most instances, if not all, these phenomena may well manifest various effects of the human hand, but surely that does not make them artifacts. More directly, nature has very much to say in the causality of these events—indeed, in the causality of virtually any major event termed "environmental." Nature's causality is of crucial importance in its own right, and any discussion of mastery only magnifies that significance.

At the point where mastery of nature arises in *The End of Nature*, McKibben articulates an idea reproducing his larger argument. McKibben unerringly spots the schools of mastery offering solutions to our ecological predicament. "By dint of our powerful intellects," he writes, "...we may have a way out. Just in time.... It's not certain that genetic engineering and macromanagement of the world's resources will provide a new cornucopia, but it certainly seems probable. We are a talented species. Why, then, does it sound so awful? Because, of course, it represents a second end of nature.... But this won't be by accident — this will be on purpose."[15] Considered generally, it is the purposeful character of human governance that may compound human artifice and give rise to a "second end of nature." Considered more specifically, it is especially the genetic reconstitution of nature that destroys any definition of nature by placing natural material entirely at the disposal of our artificial designs and purposes. McKibben concludes his discussion of mastery with the following thought:

> In such a world — shifting, without responsibility or moral center, lonely — everything will be possible, eventually including, perhaps, immortality. Why die? What good reason is there? Why age?... Some of this is speculation, certainly; no one can say with any exactness what will result from a development as awesome as the cracking of the gene. But if that technology falters, some other may emerge. It is the logical outcome of our defiant belief that we must forever dominate the world to our advantage as we have dominated it the last hundred years. If we are going to go on increasing our numbers, accumulating more possessions, using more resources, then we will have to learn new ways, and genetic engineering and macromanagement seem the most promising.[16]

While he is offended and frightened by the prospect of total mastery, McKibben readily foresees its eventuality. He does go on to call for a change of habits, which is the only option we have apart from the "logical outcome" of the will to dominate. But McKibben's fundamental conception of the end of nature draws his understanding up short of any radical criticism of mastery. If nature's loss of independence means "there is nothing but us," then nature has nothing to say over and against our "defiant belief." Questions concerning the control of causality, or, more generally, a critique of any proposals of mastery finds no expression here simply because nature has no voice. Nature does possess a voice, however, and we have tried to listen to it in this critical study of the belief in mastery. But it would be foolish to leave behind McKibben's work on this point of difference, for *The End of Nature* does directly shed light on a matter most wanting illumination.

Bill McKibben's thinking always is sensible and insightful, and the corpus of his writing about the environmental crisis over these past two decades deserves the wide acclaim it has received among ecologists, environmentalists, and others. Stealing a judgmental line from Tony Judt's assessment of another thinker in another context, we would argue that Bill McKibben "gets the big ideas right." In *The End of Nature* McKibben drives directly to the heart of things in considering what a change of habits must entail. Although he admits that "our impulse to control nature may be too strong to stop,"[17] his refusal to accept defeat points remedial thinking beyond instrumental grounds of understanding. "Should we so choose, we could exercise our reason to do what no other animal can do: we could limit ourselves voluntarily.... What a towering achievement that would be.... Such restraint — not genetic engineering or planetary management — is the real challenge, the hard thing. Of course we can splice genes. But can we *not* splice genes?"[18] The particular question of gene splicing aside, McKibben is not so naïve as to think human groups will start lining up to see which one will be the first to initiate the towering achievement of self-restraint. Given to generous understatement, he tells us at another point, "A voluntary simplification of lifestyles is not beyond our abilities, but it is probably outside our desires."[19] Once again, Bill McKibben hits the bullseye. There are those, of course, for whom will and desires completely align in the institutional wealth and power of the status quo. They are hardly likely to want any change of any kind. But McKibben's point reaches most of the rest of us. And in this larger cultural regard there appears yet another way in which belief in mastery blinds us to just what needs critical attention.

If humankind can dominate nature, what reason is there even to consider the idea of curbing our desires? According to this assumption of mastery, expanding instrumental capabilities simply will expand the material supplies demanded by our increasing material desires. The growth dynamic within such understanding is hugely problematic on its own terms, of course, as increasing instrumental capabilities expand our material desires in synthetic fashion. Our present concern, however, lies less with the manufactured character of so many of our desires than with the natural substratum of their material satisfaction. Although aspects of our material desires may be manufactured and made into commodities, the pleasure we experience in the satisfaction of those material desires springs from our somatic being.

While continually feeding and expanding our satisfactions in modern time, the project of mastery always has drawn upon and sustained itself through the appetitive organic nature within us. The lifeblood of that project has always been the pleasures we experience within ourselves, whether we have in mind direct and immediate material indulgence or, less directly but most importantly, the easing of labor and provision for leisure. From the very outset the project of mastery was therefore an ironic venture inasmuch as it fostered in material satisfaction the very desires nature underwrites within us on its own appetitive terms. Wanting to place external nature under human command to satisfy our material wants and desires, we moderns in the West have actually allowed those wants and desires to dominate our sensibilities.

Material appetites have placed nature in command, and it is time to turn the project of mastery inward. We need to direct it towards the human interior, where nature's appearance has by and large escaped the objectification of modern science. Any discussion of mastery in that realm is a matter of human will, not instrumental knowledge, and we need to be immediately clear about two points here. First of all, the human interior is dim because of its utter privacy; it does not appear in public light. For us, however, this interior has not the depth of Joseph Conrad's, given testimony in Kurtz's exclamation "The horror! The horror!" from some primordial heart of darkness. In fact, the interior concerning us here lies largely at the surface of consciousness where our willing engages our desires at some level of self-consciousness, real or potential. That engagement may be profound in itself without requiring any profound depth beneath it. Second, in speaking of mastery here, we are not out to resurrect John Calvin's Genevan Consistory, which feared all things relating to the flesh or to pleasure. We neither can nor should want to subdue the nature within us short of holding to willful account the excesses and/or indulgences that implicate our well-being either directly, respecting our bodies, or indirectly, respecting the nature surrounding us. Using words familiar to us in times of severe economic difficulty, willful mastery of our material wants and desires amounts to cutbacks, to retrenchment, to self-imposed restraint. Willing of this sort does, nevertheless, present an enormous challenge requiring a tremendous investment in the will.

Risking colossal understatement, willing curtailment of our material indulgences presents no easy task. Witness the obesity medically challenging so many today; and picture the couch potato, a contemporary social type widespread in every sense. Consider now the couch potato's dream in the face of developing medical risks: a pill to provide the benefits of exercise without physical exertion and gene intervention to rectify obesity without having to alter one's diet. This dream expresses the modern reflex to problems with nature, in this case the human body's nature, and, unsurprisingly, both instrumental fixes imagined here are being pursued today experimentally.[20] Answers to the couch potato's dream may be soon forthcoming, in which case he may laugh all the more loudly at talk of diet and exercise. But at the risk of what danger? Without rehashing unknown dangers respecting the polygenic or pleiotropic nature of much genetic interaction, or other matters of uncertainty, suffice it here simply to spotlight the obvious fact that risking any possible dangers is unnecessary: exercise in the place of a pill; diet in the place of genetic intervention. This altogether good and sensible response, healthy in every way, lies within the reach of the couch potato's will, but only if he would activate that will. Of course, none of this prompting appeals to the couch potato, which makes all the more urgent and necessary our need (his need) to see through the illusion of mastery. We must unsettle faith in instrumental fixes; we must expose dangerous uncertainty wherever it may be veiled by "generally safe for human use or intervention." At the same time, in the end, the couch potato actually is going to have little choice about what lies within the range of his indulgence. Beyond his desires as well as our own lies nature's shrinking provision.

Concern for sustainability and critiques of growth have long made the case for our needing to scale down substantially the demands we are making upon the Earth. Much if not all of that counsel has for decades sounded Cassandra-like, but now there is a conspicuous thinning of supply where once there was bounty. The sea comes to mind immediately in this regard. Instrumental alternatives to existing exploitative practices will go just so far, however sensible we become in that regard; and even apart from the altogether legitimate claims many groups of people have on a more equitable distribution of Earth's provision, a shrinking provision will mean some serious reduction in the industrial countries' metabolic demands on the Earth. Sooner or later the curtailment of those demands will become real and immediate, and if our will proves unable to rein in our excessive indulgence, later will become sooner all too quickly. Subject to forces both natural and political, we will learn first-hand that necessity does indeed drive a very hard bargain, if it is even willing to bargain at all. However strong the wind into which the call for voluntary restraint may be whistling, this call actually does blur the distinction between the real and the ideal: it is ideal in its wishfulness against the hard reality of our given habits in the short term, but in the long term voluntary restraint is realistic in posing the only good alternative we have to having our habits denied by hard and forceful circumstances.

Voluntary restraint spotlights the naked will. The modern age has largely denuded the will of cultural clothing that theretofore had helped keep our appetitive desires and material indulgences under wrap. Apart from individuals whose wealth, social position, and character allowed their desires free rein, traditional Western culture before modern time by and large helped to restrain our desire for material satisfactions. Christian values, for example, mediated economic activities, restraining money-lending, pricing, the rape of nature, and the like. So, too, the economy itself was bound up with religious and political institutions, having to await the functional differentiation that would unfetter its material engines in modern time. Prometheus traditionally had been a figure both admirable and punishable in Western mythology, but the modern age embraced what was admirable while forgetting what was punishable.

In the seventeenth century Thomas Hobbes signaled such a change of mind, adumbrating the sensibility soon to overtake large numbers of Western peoples. Assuming that human beings in their prepolitical state of nature most fear their own death, Hobbes pointed toward utilitarian interests as the springs of human action. He was not unambiguous in this regard, but he did anticipate Bentham's later utilitarianism in suggesting that each of us aversely seeks to avoid or minimize that which produces pain while, on the other hand, we each desire and appetitively pursue to the best of our attainment that which produces pleasure. People thus tend, generally and ordinarily, to define their happiness by the measure of material satisfactions, weighing the fulfillment of desires and wants over and against the suffering of privation and pain. However mistaken this measure may be, the fact is, such understanding would help increasingly to outfit the modern West in impressive material clothing while leaving the will itself quite bare as a possible bulwark against excessive or dangerous self-indulgence.

Turning the will into such a bulwark surely is a project of mastery, but one hardly resembling the project first envisioned by Bacon. The will is a phenomenon of the human interior which the Western philosophical tradition, informed most recently by German idealism, classifies as a faculty of the mind. It was Nietzsche, however, who knew the will best, and as he informs us most generally, "the will is precisely that which treats cravings as their master and appoints them to their way and measure."[21] Without moving towards philosophy here, we must note that the will is at the same time muscular, which is why we rightly speak of "exercising" the will to keep it "strong." We also *feel* the will in the experience of command: we feel the will in that rush of victory when we will to overcome some fear, just as we feel the will in that deep-seated strength when we will to seize control of our powerful urge to eat more chocolate and let the diet go to hell. Quite simply, we feel in command. And this command feels good.

> A man who *wills* commands something within himself that renders obedience, or that he believes renders obedience.... "Freedom of the will"—that is the expression for the complex state of delight of the person exercising volition, who commands and at the same time identifies himself with the executor of the order—who, as such, enjoys also the triumph over obstacles, but thinks within himself that it was really his will that overcame them. In this way the person exercising volition adds the feelings of delight of his successful executive instruments ... to his feelings of delight as commander.[22]

Nietzsche here spells out what each of us should know from personal experience. Taking a page from Pascal's aphoristic thought, the will has pleasures of its own about which the body's desires know nothing.

The will's pleasure in command is hardly inconsequential for a will standing naked in the face of the normally strong desire for material satisfactions, or, more pointedly, self-indulgence of any sort. Confronting such desire and withstanding it is anything but an easy project of mastery, even if we limit our aim to retrenchment. Some individuals are far better equipped than others to undertake such a project, but most of us likely need an exercise regime in any case. Our wills need strengthening in and for this task. At the same time, resistance to the will must be encountered for the will to experience itself, to feel itself, in willing. Overcoming some opposition is the *sine qua non* of willing, for otherwise we have no experience of command.[23] In *The Immoralist* by André Gide, the protagonist Michel finds his will "broken" precisely because its very strength has destroyed all resistance, and he suffers languor without the ability to will anything.[24] Dostoevsky's "underground man," on the other hand, inhabits a world that resists him in every way. Neurotic in the extreme and unable to function happily in the world, the underground man intentionally elicits disgust and opposition from others so as to achieve pleasure, his sole pleasure, in willing defiance of their sensibilities.[23] Whereas the absence of resistance renders Michel's will impotent, opposition itself enables the underground man to stiffen his will in a world that otherwise renders him powerless. In this very regard, then, the project of mastery we are proposing is quite unlike that of the instrumental kind which led moderns to assume that nature has nothing to say. The command we propose is ever mindful of nature's voice in the persistence and strength of those wants and desires with which we must contend. Without feeling their strength, we cannot experience our own command. But in experiencing our own command, we might find pleasure of a non–material sort in an altogether new project of mastery.

Our choosing to activate the will is the issue, of course, and McKibben surely was on the mark where he noted that while self-restraint is not beyond our ability, it surely falls somewhere outside of our desires. In this same regard McKibben also has claimed in another work, where his argument aims to preserve our natural humanity from the genetic engineers, that it is our ability to consciously limit ourselves that, in fact, makes us unique within the animal kingdom: "We can decide not to do something that we are able to do. We can set limits on our desires. We can say, 'Enough.'"[24] Quite simply, we must will to will, which is truly a catch-22 predicament if ever there was one. Yet, short of throwing up our hands in complete resignation, we must attempt to start somewhere if we are to redirect the project of mastery. The will needs whatever support it can get to help it focus on its appointed task of assuming some command.

With neither naivete nor presumption, then, we ultimately present the argument of this study in the manner of an offering, however small, in the interest of such support. Saying "enough" may become just a little less difficult — perhaps not quite so impossible — if we realize that amidst the darkness and shadows occupying the land of uncertainty, we cannot clearly see the stakes on the table or envision the risks being run when confronted by proposals to take nature under our guidance. This is the land of storytelling, a land in which the only really reasonable talk of guidance or mastery ought to respect our taking command of ourselves. And in stepping beyond any illusion of human dominion, seeking mastery of nature in a completely new key, perhaps we might find the very first pleasure of real command in our willing to will.

In Sum

Hegel saw in storytelling the mind's way of reconciling itself to a reality from which it is estranged. That much surely is true of the storytelling in these pages. This work springs primarily from the concern to understand why and how it is that we stand towards nature as we do. Towards this end our narrative and analysis have from the beginning focused on the theme of mastering nature, aiming to see how we are so incredibly powerful and yet so terribly limited, so extremely clever and yet so very blind. Essentially ignoring the powers and products of modern science and technology, we have called instrumental knowledge into serious question in a culture that locates its pride, hope, and glory in the modern *vita activa* and the crowning achievements of maker's knowledge. One would need to be deafened not to hear the Sirens' song of power sung so loudly with the instrumental backing of science and technology. The seductive lure of science and technology can be overwhelming in the extraordinarily immense and ever expanding evidence of our instrumental capabilities. But we have remained steadfast in our purpose even without any wax to stop our ears. We have remained steadfast because, in the end, we really must rethink our being with nature.

Not too long ago in discussion about the unlikelihood of our ever coming to our senses, one relatively young friend with a relatively young family declared that what humankind probably needs to open its eyes is nothing less than a massive, worldwide plague of some sort that would kill off perhaps one-third of all people. This man has a wife and one child, so his one-in-three wakeup formula would work very well in their case. He doubtlessly had not fleshed out the thinking behind his cavalier sense of what we may have to endure to experience any real enlightenment. Yet one does fear the reality behind his assessment. There is something at once terribly dangerous, remarkably silly, and altogether outrageous in our general view of nature's things and how we stand towards them.

We still gaze upon the planets and stars as did the wonderers of long ago, and wonderers still dwell among us. But they are relatively few. It is not wonder that lifts our vision today. In contemporary stargazing the planets and stars present challenges to be overcome. We turn our eyes upward as space pioneers, fully confident that we'll get there and colonize or employ whatever celestial bodies we deem useful. Like a war that can divert a population's thoughts away from grave domestic problems, this uplifting vision leaves behind the awfully frightful reality of the immediate world in which we live, which is the one and only world that truly needs our attention.

This world is the natural world that is our home. It is the only home we have. And this home is a life-world, which is much less given to our manipulative control than are the elements that make up the stars and planets. This home has been extremely hospitable to us moderns, and we, at the same time, have much to acknowledge and for which to be grateful in the accomplishments of instrumental knowledge. But the salad days of thoughtless indulgence and the confident times of fixes and solutions all lie behind us. It is time for real sobriety in our understanding of nature, and we had better get over the hangover of our promised dominion in a damned hurry. And if somehow we voluntarily can attain such sobriety, we will have achieved something really remarkably worthwhile. For then—come what may—it might truly be said with meaning far more profound than before: One small step for a man; one giant leap for mankind.

Notes

Introduction

1. In Heidegger's view, Greek philosophy lay at the root of the objectification of Being underlying our technological appropriation of things. His philosophical enterprise sought to reopen the question of Being by undercutting the entire Western metaphysical tradition. The Frankfurt School, on the other hand, saw in Greek thinking and myth the elements of rationality and mastery necessary for human emancipation from nature's grip and human oppression. For Critical Theory, however, dialectical relations bound those elements to negation in the entanglement of enlightenment with myth, reason with unreason, liberation with domination. See Max Horkheimer and Theodor W. Adorno, *Dialectic of Enlightenment*, translated by John Cumming (New York: Herder and Herder, 1972), esp. 3–80; and also Max Horkheimer, *Eclipse of Reason* (New York: Seabury Press, 1974), esp. 3–57. Coming out of this same tradition is William Leiss, *Domination of Nature* (Boston: Beacon Press, 1974).

2. George W. Fisher, "Sustainable Living: Common Ground for Geology and Theology," in *The Earth Around Us: Maintaining a Livable Planet*, edited by Jill S. Schneiderman (New York: W. H. Freeman and Company, 2000), 108–109.

3. Ficino writes of man that he "not only rules the animals by force, he also governs, keeps, and teaches them.... It is also obvious that he is the god of the elements ... [and ultimately] he is god of all materials, for he handles, changes, and shapes all of them." Marsilio Ficino, "The Soul of Man" in *The Portable Renaissance Reader*, edited by James Bruce Ross and Mary Martin McLaughlin (New York: Penguin Books, 1968), 388.

4. Making the connection between the idea of progress and Christianity, Karl Löwith states that "the irreligion of progress is still a sort of religion, derived from the Christian faith in a future goal, though substituting an indefinite and immanent *eschaton* for a definite and transcendent one." Karl Löwith, *Meaning in History* (Chicago: University of Chicago Press, 1949), 114. The historian Lynn White was much more categorical in a widely read essay where he wrote, "Our daily habits of action ... are dominated by an implicit faith in perpetual progress, which was unknown either to Greco-Roman Antiquity or to the Orient. It is rooted in, and is indefensible apart from, Judeo-Christian teleology." Lynn White, Jr., "The Historical Roots of Our Ecological Crisis" in Lynn White, Jr., *Machina Ex Deo: Essays in the Dynamism of Western Culture* (Cambridge, MA: MIT Press, 1968), 85. For the disconnection of the idea of progress from Judeo-Christian sources, see Hannah Arendt, "The Concept of History: Ancient and Modern" in Hannah Arendt, *Between Past and Future* (New York: Viking Press, 1968), 41–90.

5. Francis Bacon, *The New Organon*, edited by Fulton H. Anderson (Indianapolis: The Bobbs-Merrill Inc., 1960), 39.

6. Francis Bacon, "New Atlantis," in Francis Bacon, *Essays and New Atlantis* (Roslyn: Walter J. Black Inc., 1942), 288.

7. Richard Lewontin, *It Ain't Necessarily So: The Dream of the Human Genome and Other Illusions*, 2nd ed. (New York: The New York Review of Books, 2001), xxvii.

8. Bacon, *The New Organon*,

9. See Richard H. Popkin, *The History of Skepticism from Erasmus to Spinoza* (Berkeley: University of California Press, 1979), 1–17.

Chapter 1

1. A "crisis literature" exists on the dissolution of various forms of authority in the sixteenth and seventeenth centuries. Historians in this group tend to see the issue of crisis resolution as integral to any larger interpretation of the period. The seminal work giving rise to this literature is Paul Hazard, *The European Mind 1680–1715*, translated by J. Lewis May (Middle-

sex: Penguin Books, 1964). The original French publication appeared in 1935. For a later work that also provides a gloss of the history of this literature, see Theodore K. Rabb, *The Struggle for Stability in Early Modern Europe* (New York: Oxford University Press, 1975).

2. Popkin's fine work illuminates portions of the path we follow in this section.

3. While the present work concerns itself with the history of science, at no point does it presume itself to be any history of science. The perspective on science taken here is essentially that of cultural interpretation.

4. As will become clear in this chapter, when we speak of mastery or control of nature in the most fundamental and essential sense, it is not applied science that we have in mind.

5. John Donne, "An Anatomy of the Universe: First Anniversary" in *John Donne: The Complete English Poetry*, edited by A. J. Smith (London: Penguin Books, 1971), lines 205–213.

6. Blaise Pascal, *Pensées*, translated by A. J. Krailsheimer (Middlesex: Penguin Books, 1966), 92.

7. Ibid., 146.

8. The work of Montaigne's that is most pertinent to the following discussion is his book-length essay "An Apology for Raymond Sebond," edited by Michel de Montaigne, *The Complete Essays*, translated by M. A. Screech (London: Penguin Books, 1991), 489–683.

9. Montaigne's treatment of cultural relativity can be found ibid., 646–656.

10. Ibid., 634–635.

11. See Richard H. Popkin, *The History of Skepticism from Erasmus to Spinoza* (Berkeley: University of California Press, 1979), 46–55.

12. Montaigne, "An Apology for Raymond Sebond," 679.

13. Ibid., 680.

14. For the following discussion, see Popkin, *The History of Skepticism from Erasmus to Spinoza*, 55–86.

15. Ibid., 69.

16. Ibid., 15.

17. Pascal, *Pensees*, 154.

18. *Hegel's Lectures on the History of Philosophy*, vol. II, edited and translated by E. S. Haldane and Frances H. Simson (New York: Humanities Press, 1955), 332.

19. Two examples of this genealogical line of historical interpretation are Paolo Rossi, *Francis Bacon: From Magic to Science*, translated by Sacha Rabinovitch (Chicago: University of Chicago Press, 1968); and Frances Yates, *The Rosicrusian Enlightenment* (London: Routledge & Kegan Paul, 1972).

20. For the conventional treatment of Descartes and Bacon, see the following two standard works: Herbert Butterfield, *The Origins of Modern Science, 1300–1800*, rev. ed. (Toronto: Clarke, Irwin, & Co., 1968), 96–116; and Charles Coulston Gillispie, *The Edge of Objectivity* (Princeton: Princeton University Press, 1960), 74–95. Peter Gay does gesture in the direction of our own treatment of the two thinkers: he writes that "Bacon's and Descartes' ideas on method converged because they agreed on the true purposes of philosophy: the end of true philosophizing was mastery over nature." Peter Gay, *The Enlightenment: An Interpretation*, vol. I (New York: Alfred K Knopf, 1966), 311–312. Morris Kline takes a similar interpretive line in his *Mathematics and Western Culture* (New York: Oxford University Press, 1953), 212–213. Such gestures are meaningful, but they fall short of the most fundamental point of mastery, which respects the realm of appearance we are addressing thematically in this chapter.

21. Copernicus himself was no pyrrhonist. It was his mathematical conception of the heliostatic universe that threw into question the testimony of the senses.

22. Copernicus, "The *Commentariolus*," in *Three Copernican Treatises*, translated by Edward Rosen (New York: Dover Books, 1959), 59.

23. René Descartes, *Philosophical Essays*, translated by Laurence J. Lafleur (Indianapolis: Bobbs-Merrill Co., 1964), 76.

24. Descartes will declare that God was deeply and truly his one saving certainty right from the beginning. In fact, God is very evident at many points in Descartes's reflections. See especially his third and fourth "Meditations," in Descartes, *Philosophical Essays*, 91–118.

25. This formulation of Descartes's seems somewhat soft, considering the hard-edged ideas he has in mind. Logic and mathematics may entail clear and distinct ideas, but other kinds of thought also might be so described.

26. "Rules for the Direction of Mind," ibid., 212–213.

27. Ibid., 134.

28. As Galileo speaks of the book of the universe, "[i]t is written in the language of mathematics, and its characters are triangles, circles, and other geometric figures without which it is humanly impossible to understand a single word of it; without these, one wanders about in a dark labyrinth." Galileo Galilei, "The Assayer" in *Discoveries and Opinions of Galileo*, edited and translated by Stillman Drake (New York: Doubleday and Co., 1957), 238.

29. Descartes, *Philosophical Essays*, 86, 197.

30. There were different formulations of this doctrine. See the classic E. A. Burtt, *The Metaphysical Foundations of Modern Science* (Garden City, NY: Doubleday and Co., 1954), *passim*.

31. General accounts of the scientific revolution told from this standpoint are Burtt; E. J. Dijksterhuis, *The Mechanization of the World*

Picture, translated by C. Dikshoorn (London: Oxford University Press, 1961); and A. Rupert Hall, *From Galileo to Newton* (New York: Dover Publications, 1981). Alexander Koyré inspired many of the interpretations coming from this perspective. See Koyré's more specialized studies in his *Metaphysics and Measurement: Essays in the Scientific Revolution* (Cambridge, MA: Harvard University Press, 1969), especially chapters 1 and 2. See also chapter 4 in his *Newtonian Studies* (Chicago: University of Chicago Press, 1965).

32. See René Descartes, "Discourse on Method," in *Philosophical Works of Descartes*, vol. 1, translated by Elizabeth S. Haldane and G. R. T. Ross (Mineola, NY: Dover Publications Inc., 1955), 119.

33. Francis Bacon, *The New Organon* (Indianapolis: Bobbs-Merrill, 1960), 91.

34. See Steven Shapin, *The Scientific Revolution* (Chicago: University of Chicago Press, 1996), 19–21.

35. Bacon, *The New Organon*, 93.

36. See, for example: Gillispie, *The Edge of Objectivity*, 75–95; and Richard Tarnass, *The Passion of the Western Mind* (New York: Harmony Books, 1991), 272–281. Perhaps the best contraposition of Bacon and Descartes is in Butterfield, *The Origins of Modern Science*, 96–116.

37. Bacon, *The New Organon*, 52.

38. Ibid., 21.

39. Ibid., 66.

40. For a fine discussion of Bacon's new organon that does not miss the point, see Michel Malherbe, "Bacon's Method of Science" in *The Cambridge Companion to Bacon*, edited by Markku Peltonen (Cambridge: Cambridge University Press, 1996), 75–98.

41. Bacon, *The New Organon*, 53.

42. Ibid., 22.

43. Ibid., 24, 25.

44. Ibid., 107.

45. Ibid., 39.

46. In later chapters we will consider other kinds of causality that differ from the form of causality Bacon addresses and classical Newtonian science features. So, too, there are different kinds of experimentation in the kingdom of modern science, and the kind we are discussing here is the ideal-typical reigning authority within that kingdom. Other kinds will be addressed in later chapters.

47. Bacon, *The New Organon*, 29.

48. Antonio Pérez-Ramos, "Bacon's forms and the Maker's Knowledge Tradition," in Peltonen, *The Cambridge Companion to Bacon*, 109, 111.

49. Peter Dear, *Revolutionizing the Sciences* (Princeton: Princeton University Press, 2001), 64. See also the two Appendices, "The Nature-Art Relationship and the Machine of the World" and "Truth and Utility in the Science of Francis Bacon" in Paolo Rossi, *Philosophy, Technology and the Arts in the Early Modern Era* (New York: Harper & Row, 1970), 137–173.

50. Ibid., 95–96.

51. Ibid., 19.

52. Francis Bacon, "New Atlantis," in Francis Bacon, *Essays and New Atlantis* (Roslyn, New York: Walter J. Black, 1942), 288.

53. No one better appreciated such prescience than the gifted writer Loren Eiseley, who entitled his short work on Bacon *The Man Who Saw Through Time* (New York: Charles Scribner's Sons, 1973).

Chapter 2

1. John Henry covers this point and related material very nicely in his *The Scientific Revolution and the Origins of Modern Science* (New York: Palgrave Press, 2002), 3–5.

2. Steven Shapin is one of those who opposes the idea of a scientific revolution. He opens his book with this clever gambit: "There was no such thing as the scientific revolution, and this is a book about it." Steven Shapin, *The Scientific Revolution* (Chicago: University of Chicago Press, 1996), 1. Shapin follows a long line of critics going back to Pierre Duhem, who a century ago identified in late medieval schools various works and beliefs presaging later ideas and positions associated with those sixteenth- and seventeenth-century figures commonly called "founders" of modern science. E. J. Dijksterhuis famously broadened the scope of this continuity thesis by including the ancient Greeks in an extremely lengthy story of the mathematization of the West's *Weltanschauung*: see E. J. Dijksterhuis, *The Mechanization of the World Picture*, translated by C. Dikshoorn (London: Oxford University Press, 1961). For an overview of the issue, see also Bernard Cohen's discussion of "evolution" versus "revolution" in his *Revolution in Science* (Cambridge, MA: Harvard University Press, 1985), 561–563. Our own position in this work clearly argues for discontinuity over continuity in assuming that something in the way of a scientific revolution certainly did take place. This position keeps good company among historians of science. For a good overview of how the idea of the scientific revolution took shape historically, see David C. Lindberg, "Conceptions of the Scientific Revolution from Bacon to Butterfield: A Preliminary Sketch" in *Reappraisals of the Scientific Revolution*, edited by David C. Lindberg and Robert S. Westfall (Cambridge: Cambridge University Press, 1990), 1–26. A full account of the many perspectives and interpretive issues surrounding the rise of modern science can be found in H. Floris Cohen,

The Scientific Revolution: A Historiographical Inquiry (Chicago: University of Chicago Press, 1994).

3. G. W. F. Hegel, *Hegel's Philosophy of Right*, translated by T. M. Knox (London: Oxford University Press, 1967), 13.

4. See such widely diverse treatments as E. A. Burtt, *The Metaphysical Foundations of Modern Science* (Garden City, NY: Doubleday and Co., 1954); and I. Bernard Cohen, *The Birth of a New Physics* (Garden City, NY: Doubleday & Co., 1960). Thomas Kuhn's *The Copernican Revolution* (Cambridge, MA: Harvard University Press, 1957) focuses mainly on the Copernican conception of planetary astronomy and its impact, but he does see Newton's work as the completion of this conceptual revolution (see pages 252–265). While initiating his discussion with Nicholas of Cusa, a fifteenth-century thinker, Alexander Koyré essentially follows suit in his *From the Closed World to the Infinite Universe* (Baltimore: John Hopkins Press, 1957). All four of these works are classic interpretations, and we see the same launching point and, despite the inclusion of Einstein, the same grouping of giants in Stephen Hawking's recent *On the Shoulders of Giants: The Great Works of Physics and Astronomy* (Philadelphia: Running Press, 2002).

5. See Paolo Rossi, *Philosophy, Technology and the Arts in the Early Modern Era* (New York: Harper, 1970).

6. Giving wider scope to the reach of the *vita activa* than the instrumental knowledge with which we are concerned, Hannah Arendt sees in the valorization of labor and the technics associated with laboring, the central feature of the modern age. See especially her illuminating discussion of the shifting valuation from the *vita contemplativa* to the *vita activa* in Hannah Arendt, *The Human Condition* (Chicago: University of Chicago Press, 1958), 7–21.

7. See Rossi, *Philosophy, Technology and the Arts in the Early Modern Era*, 41–62; and Peter Dear, *Revolutionizing the Sciences* (Princeton: Princeton University Press, 2001), 49–64.

8. This distinction is W. D. Hackman's. See his "Scientific Instruments: Models of Brass and Aids to Discovery," in *The Uses of Experimentation: Studies in the Natural Sciences*, edited by David Gooding et al. (Cambridge: Cambridge University Press, 1989), 39–43.

9. See Dear's discussion: Dear, *Revolutionizing the Sciences*, 111–135; and also Richard S. Westfall, *The Construction of Modern Science: Mechanisms and Mechanics* (Cambridge: Cambridge University Press, 1971), 105–119.

10. See John Gascoigne, "A Reappraisal of the Role of the Universities in the Scientific Revolution" in Lindberg, "Conceptions of the Scientific Revolution from Bacon to Butterfield," 207–260.

11. Morris Bishop, *The Middle Ages* (Boston: Houghton Mifflin, 1968), 249–250.

12. See Dijksterhuis, *The Mechanization of the World Picture*, 233–240.

13. Westfall, *The Construction of Modern Science*, 31.

14. Gascoigne, "A Reappraisal," 217.

15. Burtt's discussion of Kepler is most illuminating. See Burtt, *The Metaphysical Foundations of Modern Science*, 63–71.

16. The interpretation of Galileo's work presented here seems to be the consensus arrived at after debate between two extremes, the one being that his math was not that important and the other that his experimental work meant little. Ernst Mach's earlier reading of Galileo as a strict empiricist fell under direct attack when Koyré put forth his widely influential reading of Galileo as a Platonist of a modern sort. See, for example, the latter's "Galileo and Plato," in Alexander Koyré, *Metaphysics and Measurement: Essays in the Scientific Revolution* (Cambridge, MA: Harvard University Press, 1969), 16–43. Following Koyre's lead, Thomas Kuhn interprets Galileo similarly, but with another argument in mind: see the "Mathematical versus Experimental Traditions in the Development of Physical Science" in his *The Essential Tradition: Select Studies in Scientific Tradition and Change* (Chicago: University of Chicago Press, 1977), 31–65. For the more balanced kind of treatment that seems now to prevail, see Dudley Shapere, *Galileo: A Philosophical Study* (Chicago: University of Chicago Press, 1974), 126–145. See also Cohen's assessment of the debate in *The Scientific Revolution*, 108–110.

17. This universality overturned the two-tiered ontology of the medieval cosmology in which imperfection (change, passing out of *Being*) governed the sublunar world while perfection (immutability, permanence) reigned above the Earth's place. Galileo's telescopic discoveries played no small role in the eventful disintegration of that ontology.

18. See Charles Coulston Gillispie, *The Edge of Objectivity* (Princeton: Princeton University Press, 1960), 50–53.

19. Scholars generally agree on this matter though, in a much earlier work, Kline directly credits Galileo with Newton's first law of motion (inertia): Morris Kline, *Mathematics and Western Culture* (New York: Oxford University Press, 1953), 190. For a rejoinder on the issue, see John Herivel, *The Background to Newton's Principia* (London: Oxford University Press, 1965), 35–53.

20. Shapere, *Galileo: A Philosophical Study*, 106.

21. A. Rupert Hall, *From Galileo to Newton* (New York: Dover, 1981), 276.

22. A recent biography titularly relies on

Keynes' words: Michael White, *Isaac Newton: The Last Sorcerer* (New York: Basic Books, 1997). For a good discussion of Newton's larger religious concerns, see Gale E. Christianson, *In the Presence of the Creator: Isaac Newton and His Times* (London: Collier Macmillan, 1984), 203–265.

23. Isaac Newton, *Principia* in Hawking, *On the Shoulders of Giants*, 1038.

24. "Cotes' Preface to the Second Edition of the *Principia*," in Isaac Newton, *Principia*. vol. I, translated by Andrew Motte revised by Florian Cajori (Berkeley: University of California Press, 1962), xvii.

25. Newton, "From a Letter to Oldenburg," in *Newton's Philosophy of Nature: Selections from His Writings*, edited by H. S. Thayer (New York: Hafner Press, 1953), 7–8.

26. We will pursue this matter in some detail in the next chapter.

27. Alexander Pope, "Intended for Sir Isaac Newton," in *Alexander Pope: Selected Works*, edited by Louis Kronenberger (New York: Modern Library, 1951), 330.

28. See Thomas Hankins, *Science and the Enlightenment* (Cambridge: Cambridge University Press, 1985), 5–16.

29. Ibid., 110. This point fails to acknowledge the fact of there being an astrophysics based upon Newtonian laws.

30. See Betty Jo Teeter Dobbs and Margaret C. Jacob, *Newton and the Culture of Newtonianism* (New York: Humanities Books, 1995), 73–74.

31. Ibid., 78.

32. Ibid.

33. Mordecai Feingold, *The Newtonian Moment: Isaac Newton and the Making of Modern Culture* (New York: Oxford University Press, 2004), 68. The following discussion of the Dutch leans heavily on Feingold's text, which was written for the Newton exhibition held at the New York Public Library, with cooperation from Cambridge University, between October 2004 and February 2005.

34. Peter Gay, *The Enlightenment: An Interpretation*, vol. II (New York: Alfred A. Knopf, 1969), 18.

35. Dobbs and Jacobs, *Newton and the Culture of Newtonianism*, 83.

36. Ibid., 107.

37. Voltaire, "Letters on the English" in *Main Currents in Western Thought*, edited by Franklin Le Van Baumer (New Haven: Yale University Press, 1978), 376.

38. See Feingold, *The Newtonian Moment*, 103–116.

39. Dobbs and Jacobs, *Newton and the Culture of Newtonianism*, 85, 88.

40. Feingold, *The Newtonian Moment*, 104.

41. Gay, *The Enlightenment*, 130.

42. See ibid., 130–133.

43. Carl L. Becker, *The Heavenly City of the Eighteenth-Century Philosophers* (New Haven: Yale University Press, 1932), 30, 31.

44. Arianism sprang from the heretical view that since God created Jesus, Jesus had lesser standing in the ranks of the divine. Newton's very personal Christianity entailed Arianism and its subversion of the Trinity.

45. Peter Gay's large two-volume work on the Enlightenment generally might be considered a good rejoinder to such reductionism.

46. David Hume, "Of Miracles," in *Enquiries Concerning the Human Understanding and Concerning the Principles of Morals*, edited by L. A. Selby-Bigge (Oxford: Oxford University Press, 1957), 109–131.

47. See Hankins, *Science and the Enlightenment*, 37–41.

48. Dobbs and Jacobs, *Newton and the Culture of Newtonianism*, 57.

49. See Justin A. I. Champion, "Deism," in *The Columbia History of Western Philosophy*, edited by Richard Popkin (New York: Columbia University Press, 1999), 437–445.

50. Gay, *The Enlightenment*, vol. I, 18.

51. This characterization is Loren Eiseley's. Eiseley, *The Man Who Saw Through Time* (New York: Charles Scribner's Sons, 1973), 36.

52. J. B. Bury, *The Idea of Progress* (New York: Dover Books, 1932), 160. This classic work remains an essential text on the subject.

53. For a good discussion of the *Encyclopedia*, see Hankins, *Science and the Enlightenment*, 163–170.

54. Denis Diderot, "*Encyclopedia*," in *The Portable Enlightenment Reader*, edited by Isaac Kramnick (New York: Penguin Books, 1995), 17–18.

55. On the eve of the Enlightenment, Fontenelle articulated his classic minimalist interpretation of progress as the historically continuing advancement of knowledge. See Bernard de Fontenelle, "Digressions on the Ancients and Moderns," in Baumer, *Main Currents in Western Thoughts*, 356–359.

56. Hankins, *Science and the Enlightenment*, 165.

57. See Jean Le Rond d'Alembert, "The Human Mind Emerged from Barbarism," in Kramnick, *The Portable Enlightenment Reader*, 7–9.

58. Gay, *The Enlightenment*, vol. II, 252–253.

59. "Jean Le Rond d'Alembert: Preliminary Discourse to the Encyclopedia of Diderot," translated by Richard Hooker, 8 June 2005, http://www.wsu.edu:8080/~wldciv/world_civ_reader/world_civ_read.

60. Franklin L. Baumer, *Modern European Thought: Continuity and Change in Ideas, 1600–1950* (New York: Macmillan, 1977), 148.

61. Montesquieu informs us in the "Preface" to this work, "I have set down the principles, and I have seen particular cases conform to them as if by themselves, the historians of all nations being but their consequences, and each particular law connecting with another law or dependent on a more general one." Baron de Montesquieu, *The Spirit of the Laws* (Cambridge: Cambridge University Press, 1989), xliii.

62. Ernst Cassirer, *The Philosophy of the Enlightenment*, translated by Fritz Koelln and James Pettegrove (Boston: Beacon Press, 1951), 214–215.

63. Joseph Priestley, "Essay on Government," in Baumer, *Main Currents in Western Thought*, 453.

64. Marquis de Condorcet, "The Progress of the Human Mind," ibid., 454.

65. Crane Brinton, *The Shaping of Modern Thought* (Englewood Cliffs, NJ: Prentice-Hall, 1963), 138.

66. For an essay finely balanced in its judgment of the French Revolution, see Robert Darnton, "The Kiss of Lamourette," in *The Kiss of Lamourette: Reflections in Cultural History* (New York: W. W. Norton, 1990), 3–20.

67. Immanuel Kant, "What Is Enlightenment?" in *Foundations of the Metaphysics of Morals and What Is Enlightenment?*, edited and translated by Lewis White Beck (Chicago: University of Chicago Press, 1950), 286.

68. Edmund Burke, *Reflections on the Revolution in France* (Indianapolis: Bobbs- Merrill Co., Inc., 1955), 110.

69. For an interesting perspective on the impact that the French Revolution and the Napoleonic wars would have on nineteenth-century historical sensibility, see Peter Fritzsche, *Stranded in the Present: Modern Time and the Melancholy of History* (Cambridge, MA: Harvard University Press, 2004). Fritzsche sees modern historical sensibility arising with the sense of irrevocable loss many Europeans experienced with the French Revolution and the wars that followed. Cut off from their past, the first true moderns obtained their historical consciousness in an orphanage, where nostalgia and melancholy marked their being in time. Fritzsche pays no attention to the abandonment of the past obtaining in the rise of modern science as his eyes seem fastened on the Romantic understanding that emerges at the beginning of the nineteenth century. Science should get some attention in that same context, however, especially in view of Romanticism's attack on materialism and the outlook behind it. Indeed, Romantic thinkers themselves saw in the nature of modern science and scientific enlightenment the cultural demise of history in the eighteenth century.

70. Shapin, *The Scientific Revolution*, 65.

71. Dobbs and Jacobs, *Newton and the Culture of Newtonianism*, 45.

72. Kline, *Mathematics and Western Culture*, 212, 213.

73. Quoted by Milič Čapek, *The Philosophical Impact of Contemporary Physics* (New York: Van Nostrand Reinhold, 1961), 122.

74. Imperfect were the data with which Newton worked, and so, accordingly, imperfect was the working mechanism of his universe. Newton's God was an active master mechanic whose intervention, where needed, saved the universe from going off kilter. Corrections in the data and mathematical advances by Laplace and Lagrange in the eighteenth century later perfected the "machine" that enlightenment thinkers saw when they looked at the universe in light of Newtonian science.

75. Čapek, *The Philosophical Impact of Contemporary Physics*, 125.

76. See Carl G. Hempel and Paul Oppenheim's highly influential essay "Studies in the Logic of Explanation," in *Readings in the Philosophy of Science*, edited by Baruch A. Brody (Englewood Cliffs, NJ: Prentice-Hall, 1970), 8–23.

Chapter 3

1. It is from the end of the Napoleonic Wars to the outbreak of World War I that John Lukacs measures the century politically. See his *The End of the Twentieth Century and the End of the Modern Age* (New York: Ticknor & Fields, 1993), 1–3.

2. See Simon Schama's discussion in *A History of Britain*, vol. III (New York: Hyperion, 2002), 144–150. That very evening Victoria recorded in her diary that the day had been "a day to live for ever." "Document 195," in *Documents of the Industrial Revolution: 1750–1850*, edited by Richard L. Tames (London: Hutchinson Educational Ltd., 1971) 185.

3. Schama, *A History of Britain*, 148.

4. Most historians see England's launching of the industrial revolution happening in the last half of the eighteenth century. Two classic works on the early phase(s) of the industrial revolution are T. S. Ashton, *The Industrial Revolution 1760–1830* (New York: Oxford University Press, 1964); and Phyllis Deane, *The First Industrial Revolution* (London: Cambridge University Press, 1967).

5. Hans Jonas, *Philosophical Essays: From Ancient Creed to Technological Man* (Chicago: University of Chicago Press, 1974), 74.

6. See Margaret Drabble, "The Great Exposition," in *Victorian England*, edited by Clarice Sinisher (San Diego: Greenhaven Press, Inc., 2000), 74–82.

7. Karl Marx and Friedrich Engels, "Mani-

festo of the Communist Party," in *Selected Works* (Moscow: Progress Publishers, 1968), 39–40.

8. Karl Marx, "Theses on Feuerbach" ibid., 30.

9. Peter Dear, *Revolutionizing the Sciences* (Princeton: Princeton University Press, 2001), 170.

10. See John U. Nef, *Western Civilization Since the Renaissance: Peace, War, Industry and the Arts* (New York: Harper and Row, 1963), 315–322.

11. See Erich Heller's illuminating "Faust's Damnation: the Morality of Knowledge," in *The Artist's Journey into the Interior and Other Essays* (London: Secker and Warburg, 1965), 3–44.

12. The immediate discussion draws heavily from Jonas, *Philosophical Essays*, 75–79.

13. Ibid., 76.

14. Ibid., 78.

15. Leiden jars had been used to store frictional electricity, with little practical payoff.

16. Lewis Mumford, *The Myth of the Machine: The Pentagon of Power* (New York: Harcourt Brace Javonovich, 1970), 121.

17. Jonas, *Philosophical Essays*, 77.

18. Alfred North Whitehead, *Science and the Modern World* (New York: The Free Press, 1953), 96–97.

19. See David Knight, *The Age of Science* (Oxford: Basil Blackwell Ltd., 1986), 52–69.

20. S. C. Burchell, *Age of Progress* (New York: Time Incorporated, 1966), 35–36.

21. David Knight's book covers these developments thoroughly.

22. Knight, *The Age of Science*, 170.

23. See John Gribben, *The Scientists* (New York: Random House, 2002), 410–412.

24. See Gribben's discussion ibid., 381–384.

25. Ibid., 381.

26. Hobsbawm refers to the "dual revolution" between 1789 and 1848 as constituting the greatest historical transformation since the invention of agriculture and metallurgy. E. J. Hobsbawm, *The Age of Revolution: 1789–1848* (New York: Mentor, 1962), 17.

27. Lewis Mumford, *The Myth of the Machine: Technics and Human Development* (New York: Harcourt, Brace & World, Inc., 1967), 294.

28. This term *neotechnical* comes from an earlier work of Mumford's that traced the development of machinery and its changing social and cultural impact. See Mumford, *The Myth of the Machine: Technics and Civilization*, especially 212–267.

29. H. G. Alexander (ed.), *The Leibniz-Clarke Correspondence* (New York: Manchester University Press, 1956). Alexander Koyré provides a lucid discussion of the issues dividing Newton and Leibniz in his *From the Closed World to the Infinite Universe* (Baltimore: Johns Hopkins Press, 1957), 235–272.

30. See E. A. Burtt, *The Metaphysical Foundations of Modern Science* (Garden City, NY: Doubleday and Co., 1954), 244–264.

31. Max Jammer, *Concepts of Space: The History of Theories of Space in Physics*, 2nd ed. (Cambridge: Harvard University Press, 1969), 116.

32. Leibniz wrote, "If the means, which causes an attraction properly so-called, be constant, and at the same time inexplicable by the powers of creatures [i.e. things], and yet be true; it must be a perpetual miracle: and if it is not miraculous, it is false. 'Tis a chimerical thing, a scholastic occult quality." "Leibniz's Fifth Paper" in Alexander, *The Leibniz-Clarke Correspondence*, 94.

33. In the "General Scholium" ending Book III of the *Principia*, Newton wrote with reference to the cause of gravity: "I have not been able to discover the cause of those properties of gravity from phaenomena, and I frame no hypotheses; for whatever is not deduced from the phaenomena, is to be called an hypothesis; and hypotheses, whether metaphysical or physical, whether of occult qualities or mechanical, have no place in experimental philosophy." Isaac Newton, "*Principia*," in *On the Shoulders of Giants: The Great Works of Physics and Astronomy*, edited by Stephen Hawking (Philadelphia: Running Press, 2002), 1159. Given Newton's somewhat pejorative use of the word "hypothesis," Koyré's translation of "*hypotheses non fingo*" as "I feign no hypotheses" seems altogether apt, which is why we employed his "feign" rather than "frame" in our own text. See Koyré's discussion of "*hypotheses non fingo*" in his *Newtonian Studies* (Chicago: University of Chicago Press, 1965), 35–52.

34. One such instance was further on in the same "General Scolium" noted above.

35. H. G. Alexander, "Introduction," *The Leibniz-Clarke Correspondence*, xxxiii. Koyré puts the same point more baldly where he declares, "Eighteenth-century thought became reconciled to the ununderstandable—with very few exceptions." Koyré, *Newtonian Studies*, 163.

36. Albert Einstein, "Autobiographical Notes," in *Albert Einstein: Philosopher-Scientist*, edited by Paul Arthur Schilpp (Evanston: The Library of Living Philosophers Inc., 1949), 19, 25.

37. The Dutch Cartesian Christiaan Huygens had in Newton's time proposed the idea that light is an undulatory wave phenomenon. Young, on the other hand, did properly understand light to be a transverse wave, and Gribben writes that after Young's idea had been corroborated by both Augustin Fresnel's experimental work and performed under the auspices of the French Academy in 1817, "the wave model of light had to be upgraded to a theory from a hypothesis." Gribben, *The Scientists*, 409. For the

larger discussion here, see Mary Hesse, *Forces and Fields: A Study of Action at a Distance in the History of Physics* (Totowa, NJ: Littlefield, Adams & Co., 1965), 189–225; and also Albert Einstein and Leopold Infeld, *The Evolution of Physics: From Early Concepts to Relativity and Quanta* (New York: Simon and Schuster, 1938), 125–153.

38. Joseph Agassi refers to Farraday's improbable success as a "Cindarella story." See his delightful *The Continuing Revolution: A History of Physics from the Greeks to Einstein* (New York: McGraw-Hill, 1968), 197–203.

39. Charles Coulston Gillispie, *The Edge of Objectivity* (Princeton: Princeton University Press, 1960), 459.

40. Norwood Hanson quotes Helmholtz's offering the following prescription: "To understand a phenomenon means nothing else than to reduce it to the Newtonian laws. Then the necessity for explanation has been satisfied in a palpable way." Norwood Russell Hanson, *Patterns of Discovery* (London: Cambridge University Press, 1965), 91.

41. See Gribben's discussion of Maxwell in Gribben, *The Scientists*, 425–433.

42. Einstein and Infeld, *The Evolution of Physics*, 146.

43. Jammer, *Concepts of Space*, 144.

44. Einstein, "Autobiographical Notes," 25–27, 31–33.

45. Einstein's "Special Theory" appeared in papers Einstein published in 1905, and the "General Theory" followed a decade later in 1915. Addressing the Michelson-Morley findings head on, the Special Theory dismissed the ether while relativizing space, time, and motion in the idealized case of inertial systems, which are observational frames of reference moving or appearing to move in uniform rectilinear motion relative to one another. Mendel Sachs offers a splendid illustration: Imagine a universe completely empty but for a single particle that is moving through space in uniform rectilinear motion (pure inertial motion). Personify both that particle and the universe and realize now that the description just given comes from the universe, which believes itself to be at rest and accordingly attributes motion to the particle. From the particle's standpoint, it is itself at rest, and motion belongs to the universe moving by it. Mendel Sachs, *Einstein versus Bohr* (La Salle: Open Court Publishing Co., 1988), 182–183. Einstein's full exposition of the Special Theory is in his *Relativity: The Special and the General Theory*, translated by Robert W. Lawson (London: Methuen & Co. Ltd., 1954), 1–57.

46. The General Theory looks beyond the Special Theory, to accelerated motion and, most significantly, gravitational fields. Ibid., 59–114.

47. Einstein and Infeld, *The Evolution of Physics*, 244.

48. Ibid., 242.

49. Einstein illustrates this equivalence in one of the *Gedenkenexperimenten* that he offers in *The Evolution of Physics*, 214–222. Imagine, if you will, one observer in an elevator and another observer poised freely in space beyond the elevator. This latter fellow, the outside observer, sees that the elevator is attached to a rope that pulls the elevator upward. For the outside observer, accelerated motion explains why something dropped inside the elevator falls to the floor. For the fellow inside the elevator, the inside observer, there is a different explanation. He assumes to all intents and purposes that his elevator is at rest, and he attributes things' falling to the floor to the gravitational field in which he and his elevator are situated. In this case Einstein wants us to privilege the inside observer's understanding inasmuch as he seeks to turn the point of equivalence towards a field explanation of motion itself.

50. Einstein wrote: "The problem is to set up a relativistic theory for the total field. The most important clue to its solution is that there exists already the solution for the special case of the pure gravitational field. The theory we are looking for must therefore be a generalization of the theory of the gravitational field." Albert Einstein, "On the Generalized Theory," *Scientific American Reprint*, no. 209 (April 1950): 6.

51. See I. Bernard Cohen, *Revolution in Science* (Cambridge, MA: Harvard University Press, 1985), 414–417. For Einstein's own highly technical discussion, see "Appendix Three" in his *Relativity*, 123–132.

52. For a sample of some of the debates in action, see *Minnesota Studies in the Philosophy of Science: Scientific Explanation, Space, and Time*, vol. III, edited by Herbert Feigle and Grover Maxwell (Minneapolis: University of Minnesota Press, 1962), 3–357 *passim*. A summary overview of some of the major positions is J. O. Wisdom, "Four Contemporary Interpretations of the Nature of Science," *Foundations of Science*, vol. 1, no. 3 (1971).

53. For a good description of this school, see Avrum Stroll, "Logical Positivism," in *The Columbia History of Western Philosophy*, edited by Richard H. Popkin (New York: Columbia University Press, 1999), 621–629.

54. Popper's interesting reflection on the origins of his philosophical perspective opens his *Conjectures and Refutations: The Growth of Scientific Knowledge* (New York: Harper & Row, 1965), 3–65. See also Karl R. Popper, *The Logic of Scientific Discovery* (Toronto: University of Toronto Press, 1959), 27–92.

55. See Thomas S. Kuhn, *The Structure of Scientific Revolutions*, 2nd ed. (Chicago: Univer-

sity of Chicago Press, 1970). Kuhn's thinking later would take a turn towards developmental change, following the evolutionary idea of organic change and biological growth. His early work nevertheless "counts," as they say. For a look at his later thinking, see his *The Road Since Structure* (Chicago: University of Chicago Press, 2000).

56. Popper chaired a symposium on Kuhn's work in the mid-1960s, and out of that conference came *Criticism and the Growth of Knowledge*, edited by Imre Lakatos and Alan Musgrave (London: Cambridge University Press, 1970). A more recent work bringing the two thinkers together is Steve Fuller, *Kuhn vs. Popper: The Struggle for the Soul of Science* (New York: Columbia University Press, 2004).

57. Popper, *Conjectures and Refutations*, 51.

58. F. S. C. Northrop, "Einstein's Conception of Science," in Schilpp, *Albert Einstein*, 407.

59. Much of the work leading up to the Solvay Conference already was quite revolutionary. See Edmund Blair Bolles, *Einstein Defiant: Genius versus Genius in the Quantum Revolution* (Washington DC: Joseph Henry Press, 2004). Two books appearing in 2007 suggest the magnitude of the quantum upheaval in subtitles possessing close similarity: David Lindley, *Uncertainty: Einstein, Heisenberg, Bohr, and the Struggle for the Soul of Science* (New York: Doubleday, 2007); and Gino Segrè, *Faust in Copenhagen: A Struggle for the Soul of Physics* (New York: Penguin Books, 2007). Segrè is especially good on the personalities and relationships of the Copenhagen folks.

60. For the following background discussion see Karl Darrow, "The Quantum Theory," *Scientific American Reprint*, no. 205 (March 1952); and also Werner Heisenberg, *Physics and Philosophy: The Revolution in Modern Physics* (New York: Harper & Row, 1958), 27–43.

61. Werner Heisenberg, *Physics and Beyond: Encounters and Conversations*, translated by Arnold Pomerans (New York: Harper & Row, 1971), 78.

62. Niels Bohr, "Discussion with Einstein on Epistemological Problems in Atomic Physics," in Schilpp, *Albert Einstein*, 237–238.

63. Heisenberg, *Physics and Beyond*,122.

64. Bohr broadly defines "complementary" or "complementarity" "in the sense that any given application of classical concepts precludes the simultaneous use of other classical concepts which in a different connection are equally necessary for the elucidation of the phenomena." Niels Bohr, *The Philosophical Writings of Niels Bohr: vol. I, Atomic Theory and the Description of Nature* (Woodbridge: Ox Bow Press, 1987), 10.

65. Quoted in Hannah Arendt, *The Human Condition* (Chicago: University of Chicago Press, 1958), 288.

66. There should be no temptation to compare this situation to Kant's denying knowledge any access to the *noumenal*.

67. Heisenberg, *Physics and Philosophy*, 45, 50.

68. Ibid., 45, 46.

69. Ibid., 54.

70. Bohr, *The Philosophical Writings of Niels Bohr*, 54–55.

71. Edmund Blair Bolles considers this distinction from yet another angle. Bolles, *Einstein Defiant*, 257–258.

72. Whitehead, *Science and the Modern World*, 34.

73. Friedrich Nietzsche, "The Birth of Tragedy," in *The Birth of Tragedy and the Case of Wagner*, translated by Walter Kaufmann (New York: Random House, 1967), 97–98.

74. Heisenberg, *Physics and Beyond*, 79, 80–81.

75. Bohr, *The Philosophical Writings of Niels Bohr*, 115.

76. Albert Einstein, et al., "Can Quantum-Mechanical Description of Nature be Considered Complete?" *Physical Review* v. 47 (May 15, 1935): 777.

77. Niels Bohr, "Can Quantum-Mechanical Description of Physical Reality be Considered Complete?" *Physical Review*. v. 48. (Oct. 15, 1935): 697.

78. Norwood Russell Hanson, "The Copenhagen Interpretation of Quantum Theory," in *Philosophy of Science*, edited by Arthur Danto and Sidney Morgenbesser (Cleveland, Ohio: Meridian Books, 1960), 457.

79. Paul K. Feyerabend, "On a Recent Critique of Complementarity: Part II," *Philosophy of Science* vol. xxxvi, no. 1 (March 1969): 92–93.

80. Bohr, *The Philosophical Writings of Niels Bohr*, 68.

81. J. Robert Oppenheimer, *Science and the Common Understanding* (New York: Simon & Schuster, 1954), 61–62.

82. Louis de Broglie, *The Revolution in Physics*, translated by Ralph Niemeyer (New York: Noonday Press, 1953), 17–18.

83. Brian Davies seems to paraphrase Heisenberg where he outlines the three stages of applying quantum theory in a typical laboratory setting. See Brian Davies, *Science in the Looking Glass: What Do Scientists Really Know?* (Oxford: Oxford University Press, 2003), 192–193.

84. Hanson, "The Copenhagen Interpretation," 463.

85. Ibid.

86. Physics teacher Tom Medlock pointed out to me that Google would not exist without quantum theory. The control of information today may go a long way towards controlling human beings in both a political and psychological sense, but even if we were to grant that

for argument's sake, such control is indeed a long way removed from the control of nature.

Chapter 4

1. Brian Davies, *Science in the Looking Glass: What Do Scientists Really Know?* (Oxford: Oxford University Press, 2003), 192.

2. J. J. C. Smart, *Between Science and Philosophy: An Introduction to the Philosophy of Science* (New York: Random House, 1968), 138.

3. It was Robert Oppenheimer's teacher Percy Bridgman who coined the term "operationism" in the mid-1920s.

4. Smart, *Between Science and Philosophy*, 151. We should also note that there are several logical critiques of instrumentalism and operationism. See, for example, Dudley Shapere, "Introduction," to *Philosophical Problems of Natural Science*, edited by Dudley Shapere (London: Macmillan, 1965), 5–15; as well as Carl Hempel, "A Logical Critical Appraisal of Operationism," in *Readings in the Philosophy of Science*, edited by Baruch A. Brody (Englewood Cliffs, NJ: Prentice-Hall, 1970), 200–210.

5. Karl R. Popper, *Conjectures and Refutations: The Growth of Scientific Knowledge* (New York: Harper & Row, 1965), 114.

6. See, for example, David Deutsch, *The Fabric of Reality: The Science of Parallel Universes and Its Implications* (New York: Penguin Books, 1997). Deutsch writes, "[P]ragmatic instrumentalism itself became a 'paradigm' which physicists adopted to replace the classical idea of an objective reality. But this is not the sort of paradigm that one understands the world through." Deutsch, *The Fabric of Reality*, 309.

7. John Horgan, *The End of Science: Facing the Limits of Knowledge in the Twilight of the Scientific Age* (New York: Broadway Books, 1997), 65, 91. Horgan, a former senior writer for *Scientific American*, argues more generally that science has essentially reached its end by virtue of its very success. Having already made the "big" discoveries, science has largely laid out the map of nature, and at this point the finitude of what is knowable makes for diminishing returns. Horgan sees the increasing costs of science and the public's loss of enchantment with science adding to the quandary of contemporary science, or at least of that science that continues to seek answers to the big questions.

8. J. O. Wisdom. "Four Contemporary Interpretations of the Nature of Science," *Foundations of Science*, vol. 1, no. 3 (1971): 271.

9. Edmund Blair Bolles, *Einstein Defiant: Genius versus Genius in the Quantum Revolution* (Washington DC: Joseph Henry Press, 2004), 303.

10. David Lindley, *Uncertainty: Einstein, Heisenberg, Bohr, and the Struggle for the Soul of Science* (New York: Doubleday, 2007), 209.

11. "Big Science," Encyclopedia Britannica Online, 10 Jan. 2008, http://www.britannica.com/bps/topic/54995/Big-Science.

12. See Gino Segrè, *Faust in Copenhagen: A Struggle for the Soul of Physics* (New York: Penguin Books, 2007), 241–243.

13. See Alvin Weinberg, "Impact of Large-Scale Science on the United States," *Science* vol. 134 (21 July 1961): 161–164.

14. See Joel Garreau's discussion of this frightening yet fascinating agency in his *Radical Evolution* (New York: Doubleday, 2005), 17–44.

15. Richard Lewontin, in *It Ain't Necessarily So*, brings together a number of articles that he wrote over the years for the *New York Review of Books*. In his discussion of the Human Genome Project, he illuminates how hype and profit can easily connect with sincere hope and an interest in knowledge. Richard Lewontin, *It Ain't Necessarily So: The Dream of the Human Genome and Other Illusions*, 2nd ed. (New York: The New York Review of Books, 2001), 135–195. Lewontin's work informs our own discussion.

16. Research entailed enormously high costs involving large teams of scientists, but before long the publicly funded consortium of scientists led by James Watson of double helix fame found itself in competition with Celera Genomics, a private company led by geneticist Craig Venter. Venter saw a quick, less expensive shortcut in the research, enabling him to compete in the race to complete the genomic map. In the end Venter's private company and Watson's consortium by political arrangement crossed the finish line together, holding hands while voicing the same promises of unprecedented medical benefits to be provided humankind.

17. Weinberg, "Impact of Large-Scale Science on the United States," 164.

18. Lewontin, *It Ain't Necessarily So*, 162–163.

19. Steven Shapin, *The Scientific Life: A Moral History of a Late Modern Vocation* (Chicago: University of Chicago Press, 2008), 295.

20. Jacob Bronowski, *The Common Sense of Science* (NY: Vintage Books, 1950), 95–96.

21. Ibid., 94.

22. Cobbled together from Bronowski's larger discussion: ibid., 70, 71, 78, 87.

23. Ibid., 104–105, 105–106.

24. Ibid., 81.

25. Ibid., 113. I articulated this idea of "instrumental adaptation" in my unpublished Ph.D dissertation. See Robert M. Tostevin, "Living without the Mastery of Nature: Beyond the Spell of Enlightenment," Dept. of Political Science, York University, Downsview, Ontario, 1977.

26. George Gaylord Simpson, *This View of*

Life: the World of an Evolutionist (New York: Harcourt, Brace & World, 1964), 99. This work was first published in 1947.

27. "Science," in *The International Encyclopedia of the Social Sciences* (New York: Crowell, Collier and Macmillan, 1968), 84, 85.

28. "Goals of Science," *The International Encyclopedia of the Social Sciences*, 3 Dec. 2005, http://en.wikipedia.org/wiki/science.

29. Aristotle's *De Anima* distinguished all living things as having one kind of soul or another.

30. See Barry Commoner, "In Defense of Biology," *Science* vol. 133 (2 June 1961): 1745–1748.

31. Ibid., 1746.

32. Lewontin, *It Ain't Necessarily So*, 277–278.

33. See Franklin Baumer's discussion in his *Modern European Thought: Continuity and Change in Ideas, 1600–1950* (New York: Macmillan, 1977), 305–308.

34. See Avrum Stroll, "Logical Positivism," in *The Columbia History of Western Philosophy*, edited by Richard H. Popkin (New York: Columbia University Press, 1999), 621–629.

35. For a detailed discussion of the shoulders on which Darwin stood, see John C. Greene, *The Death of Adam* (Ames, IA: Iowa State University Press, 1959). Also very good on this material is Loren Eiseley, *Darwin's Century: Evolution and the Men Who Discovered It* (Garden City, NY: Doubleday & Company, Inc., 1961).

36. See George Gaylord Simpson, *The Major Features of Evolution* (New York: Simon and Schuster, 1953), 58–159.

37. Charles Coulston Gillispie, *The Edge of Objectivity* (Princeton: Princeton University Press, 1960), 338.

38. It was "in the air" according to Milton Millhauser. See his piece by that title in *Darwin*, edited by Philip Appleman (New York: W. W. Norton & Co., 1970), 36–40. Well known, of course, is the fact that Alfred Wallace had reached a virtually identical understanding and that it was only Darwin's desire for paternity rights that, upon others' urging, led him to publish.

39. A good discussion of this essentialism is in Ernst Mayr, *What Evolution Is* (New York: Basic Books, 2001), 74–82.

40. Charles Darwin, *On the Origin of Species* (Cambridge, MA: Harvard University Press, 1964), 80–81.

41. See "Variational Selection," in Mayr, *What Evolution Is*, 83–114.

42. Lewontin calculates that a typical gene possessing 10,000 basic units would, given the four nucleotides, yield a number of possible DNA messages equal to 1 followed by 6,020 zeroes. Lewontin, *It Ain't Necessarily So*, 140.

43. Ibid., 70–71.

44. This fact hardly settles the long-standing issue of nature vs. nurture, though strict determinists might want it so.

45. Mayr formally defines the phenotype as the "result" of that interaction. See Mayr, *What Evolution Is*, 289.

46. Lewontin, *It Ain't Necessarily So*, 73.

47. Lamarck's understanding of evolution as the heritability of acquired traits had in some sense imprinted the phenotype on the genotype. This view has long been dismissed in the belief that genetic material cannot be transmitted from proteins to nucleic acids. Biology teacher Kathy Ketchum has pointed out that some current work in epigenetics overturns such absolute dismissal. There are exceptional cases where proteins are contributing to the otherwise "unalloyed" DNA.

48. Simpson, *The Major Features of Evolution*, 160.

49. Mayr, *What Evolution Is*, 83, 84.

50. Darwin, *On the Origin of Species*, 84.

51. Mayr, *What Evolution Is*, 7.

52. Ernst Mayr, "Methodological Issues in Biology," in *Man and Nature: Philosophical Issues in Biology*, edited by Ronald Munson (New York: Delta Books, 1971), 103.

53. Simpson, *This View of Life*, 146.

54. T. A. Goudge, *The Ascent of Life* (Toronto: University of Toronto Press, 1961), 75.

55. Michael Scriven, "Explanation and Prediction in Evolutionary Theory," in Munson, *Man and Nature*, 213. This article originally appeared in the August 28, 1959, issue of *Science*.

56. See Stephen Toulmin, *Foresight and Understanding* (New York: Harper & Row, 1961), 23–36.

57. See Niles Eldredge, *The Pattern of Evolution* (New York: W. H. Freeman and Company, 2000), 73–76. See also Eldredge's interesting note 13 on page 186.

58. Goudge, *The Ascent of Life*, 126.

59. R. C. Lewontin, *Biology as Ideology* (New York: Harper Collins Pub., 1991), 26. This volume contains the talks Lewontin gave in the 1990 Massey lecture Series over the Canadian Broadcasting System.

60. Simpson, *This View of Life*, 188, 139.

61. Stephen Jay Gould, *Wonderful Life: The Burgess Shale and the Nature of History* (New York: W. W. Norton & Co., 1989), 290. See also Gould's attack on the second-class citizenship accorded the historical sciences on 277–279.

62. Goudge, *The Ascent of Life*, 175.

63. Co-founder of string field theory, futurist, and popular lecturer Michel Kaku employs Freeman Dyson's typology of civilizations, which is based on energy sources and how they are utilized, to sketch the future of humankind's conquest of Earth and space. In this century we will lay the foundation for Type I civilization, leading to our controlling all terrestrial sources

of energy. Achieving Type II civilization, which means directly mastering the sun's energy, will take perhaps another 800 years, positioning us then to move on to the harnessing of entire star systems and establishing a truly galactic civilization millennia down the road, or time-warp. See Michel Kaku, *Visions: How Science Will Revolutionize the 21st Century* (New York: Anchor Books, 1997).

Chapter 5

1. James Watson, *The Double Helix* (New York: Atheneum, 1968), 197.
2. It should be noted that Gilgamesh, child of a goddess and a mortal man, learns that he must abandon his quest. See David Ferry's versified rendering *Gilgamesh* (New York: Farrar, Straus and Giroux, 1992).
3. Nicholas Wade, *Life Script* (New York: Simon and Schuster, 2001), 144.
4. A fine survey of this work can be found in John C. Avise, *The Hope, Hype, and Reality of Genetic Engineering* (Oxford: Oxford University Press, 2004).
5. Quoted by Garreau in Joel Garreau, *Radical Evolution* (New York: Doubleday, 2005), 252.
6. See Jeremy Rifkin's discussion in his *The Biotech Century: Harnessing the Gene and Remaking the World* (New York: Penguin Putnam, 1998), 32–36.
7. Lee M. Silver, *Challenging Nature: The Clash between Biotechnology and Spirituality* (New York: Harper Collins, 2006), xvi.
8. Michael B. Fossel, *Cells, Aging, and Human Disease* (Oxford: Oxford University Press, 2004), 290.
9. Ramez Naam, *More than Human: Embracing the Promise of Biological Enhancement* (New York: Broadway Books, 2005), 58.
10. Silver, *Challenging Nature*, xi.
11. Naam, *More than Human*, 147.
12. Gregory Stock, *Redesigning Human Beings: Choosing Our Genes, Changing Our Future* (Boston: Houghton Mifflin, 2003), 128.
13. Leonard Guarente and Cynthia Kenyon, "Genetic Pathways that Regulate Ageing in Model Organisms," *Nature* 408 (Nov. 9, 2000): 261.
14. Wade, *Life Script*, 167–168.
15. Lee M. Silver, *Remaking Eden: How Genetic Engineering and Cloning Will Alter the American Family* (New York: Harper Collins, 1998), 268.
16. Stock, *Redesigning Human Beings*, 179.
17. Naam, *More than Human*, 9.
18. Silver, *Remaking Eden*, 87.
19. Naam, *More than Human*, 127–134.
20. Ronald S. Green, *Babies by Design: The Ethics of Genetic Choice* (New Haven: Yale University Press, 2007), 34–39.
21. See ibid., 32–37.
22. Stock, *Redesigning Human Beings*, 65–72.
23. Nelson A. Wivel and LeRoy Walters, "Germ-line Gene Modification and Disease Prevention: Some Medical and Ethical Perspectives," *Science* 262 (22 Oct. 1993): 537.
24. Green, *Babies by Design*, 87.
25. See Naam, *More than Human*, 228–229.
26. David Ewing Duncan, *The Geneticist Who Played Hoops with My DNA* (New York: Harper Collins, 2005), 154.
27. Wade, *Life Script*, 178.
28. Green, *Babies by Design*, 223.
29. Stock, *Redesigning Human Beings*, 139, 143.
30. Green, *Babies by Design*, 38.
31. Gary Cziko, *Without Miracles: Universal Selection Theory and the Second Darwinian Revolution* (Cambridge, MA: MIT Press, 1995), 277.
32. Green, *Babies by Design*, 11, 15. We should note in this general context the difficulty posed by sheer numbers for the idea of directed evolution. Using population numbers at the turn of the millennium, Jon W. Gordon calculated that if 1000 engineered human reproductions were accomplished per year, they would represent but only 1/132,000 of all live births. In short, "naturally" random genetic mixes would absolutely swamp the "directed" portion of the total human gene pool. Gordon remarks additionally that genetic enhancement would itself be no guarantee of greater biological fitness, leading him to conclude that, given these two factors, "neither gene transfer nor any of the other emerging reproductive technologies will ever have a significant impact on human evolution." Jon W. Gordon, "Genetic Enhancement in Humans," *Science* 283 (26 March 1999): 2024.
33. Naam, *More than Human*, 232–233, 233–234.
34. Rifkin, *The Biotech Century*, 226.
35. Silver, *Remaking Eden*, 293.
36. Stock, *Redesigning Human Beings*, 173.
37. Quoted by Evelyn Fox Keller in her "Nature, Nurture, and the Genome Project," in *Scientific and Social Issues in the Human Genome Project*, edited by Daniel J. Kevles and Leroy Hood (Cambridge, MA: Harvard University Press, 1992), 288.
38. See Wade's discussion of this gathering in his *Life Script*, 13–22.
39. Keller, "Nature, Nurture, and the Genome Project," 293.
40. For this story and much of what follows below, see Stuart B. Levy, *The Antibiotic Paradox: How the Misuse of Antibiotics Destroys their Curative Powers*, 2nd ed. (New York: Perseus Pub., 2002). One of the world's leading experts on antibiotics, Levy leads the struggle against the overuse and misuse of antibiotics.

41. Ibid., 11.
42. See ibid., 76–79.
43. Ibid., 78–79.
44. Ibid., 228–229.
45. Ibid., 212–215.
46. Ibid., 56.
47. "The Problem of Antibiotic Resistance," *NIAID Factsheet*, April 2006, 1 of 6, http://www.niaid.nih.gov/fact-sheet/antimicro.htm.
48. "Tuberculosis," *WHO Factsheet*, March 2007, 1 of 4, http://www.who.int/mediacentre/factsheets/fs104/en/index.html.
49. Ibid.
50. Levy, *The Antibiotic Paradox*, 45–46.
51. "The Global MDR-TB and XDR-TB Response Plan," *World Health Organization*, 2007, 1 of 2, http://www.who.int/tb/features_archive/-Global_response_plan/ en/index.html.
52. Vikki Valentine, "Why TB Remains a Modern and Deadly Problem," 19 Sept. 2007, 2 of 6, html://www.npr.org/templates/story/story.php?storyId= 10551019.
53. "The Global MDR-TB, and XDR-TB Response Plan," 2.
54. "How Common are Drug-Resistant Diseases?" *National Public Radio*, 1 Sept. 2007, http://www.npr.org/templates/story/story.php?storyId=10613348.
55. Levy, *The Antibiotic Paradox*, 45.
56. See Levy's extended discussion ibid., 149–314.
57. Ibid., 292.
58. See Gautam Dantas, et al., "Bacteria Subsisting on Antibiotics," *Science* 320 (4 April 2008): 103.
59. Michael B. A. Oldstone, *Viruses, Plagues, and History* (Oxford: Oxford University Press 1998), 9.
60. Ibid.
61. See Elinor Levy and Mark Fischetti, *The New Killer Diseases* (New York: Crown Pub., 2003), 198–207.
62. "Types of Influenza Virus," *Center for Disease Control and Prevention*. 1 Oct 2007, 3 of 3, http://www.cdc.gov/flu/aout//fluviruses.htm.
63. Ibid., 2 of 3.
64. Ibid.
65. See John M. Barry, *The Great Influenza: The Story of the Deadliest Pandemic in History* (London: Penguin, 2005), 453–458.
66. On this controversy, see Gina Kolata, *Flu* (New York: Simon and Schuster, 2001), 281–306.
67. Barry, *The Great Influenza*, 454.
68. Paul M. Offit, *Vaccinated: One Man's Quest to Defeat the World's Deadliest Diseases* (New York: Harper Collins Pub., 2007), 17–19.
69. Quoted by Wendy Orent in her "What H1N1 Taught Us," 24 March 2010, 1 of 1, http://articles.latimes.com/2010/feb/26/opinion/la-oe-orent26–2010feb26.
70. Declan Butler, "Flu Officials Pull Back from Raising Global Alert Level," *Nature* 436 (7 July 2005): 6–7.
71. For a discussion of this fascinating story, see R. S. Bray, *Armies of Pestilence: The Impact of Disease on History* (New York: Barnes & Noble Books, 1996), 202–207.
72. Peter Palese of Mount Sinai and Paul Offit of the University of Pennsylvania's School of Medicine have argued categorically that H-5 viruses cannot yield an infection transmissible from human to human. See Declan Butler, "Yes, but Will It Jump?" *Nature* 439 (12 Jan. 2006): 124–125. Meanwhile, in 2007 the Fred Hutchinson Cancer Research Center statistically confirmed a 2006 Indonesian study reporting human-to-human transmission. See *Science Daily*, 2 Oct. 2007, 1 of 4. http://sciencedaily.com/releases/2007/20080 70828154944 htm.
73. Much of the following discussion relies on Levy and Fischetti, *The New Killer Diseases*, 1–31.
74. "Severe Acute Respiratory Syndrome (SARS)," *CDC Factsheet*, 25 Sept. 2007, 2 of 2, http://www.cec.gov/ncidod-/sars/faq.htm.
75. Levy and Fischetti, *The New Killer Diseases*, 22.
76. Lisa Greene, "SARS: A Learning Experience, a Lurking Fear," *St. Petersburg Times*, 31 January 2005, 1–3 of 4, http://www.sptimes.com/2005/01/31/Worldandnation/SARS-A_learning_expe/shtml.
77. Lauren J. Stockman et al., "SARS: Systematic Review of Treatment Effects," 8 Oct. 2007, 3 of 13, http://medicine.plosjournals.org/perlserev/?request =get-document&doi=10.1371/journal.pm.
78. See Greene, "SARS: A Learning Experience, a Lurking Fear," 3 of 4.
79. Quoted ibid., 2 of 4.
80. Levy and Fischetti, *The New Killer Diseases*, 21.
81. According to Sharp and Hahn the rate of HIV's evolution is "up to a million times faster than that of animal DNA." Paul M. Sharp and Beatrice H. Hahn, "Prehistory of HIV1," *Nature* 455 (2 October 2008): 605. A good overview of HIV is provided by Pete Moore in his *Super Bugs: Rogue Diseases of the Twenty-first Century* (London: Carlton Books, 2001), 120–139.
82. A highly detailed story of the investigation of HIV and the controversies infecting the understanding of its nature and origins can be found in Laurie Garrett, *The Coming Plague: Newly Emerging Diseases in a World Out of Balance* (New York: Penguin, 1994), 281–389.
83. See Annabel Kanahu and Jenni Fredriksson, "History of AIDS up to 1986," 6 May 2008, 3 of 3, http://www.avert.org/his81_86.htm.
84. Ibid.
85. Feng Gao et al., "Origin of HIV-1 in the

Chimpanzee Pan Troglodytes," *Nature* 397 (4 Feb. 1999): 436–441.

86. Elizabeth Bailes et al., "Hybrid Origin of SIV in Chimpanzees," *Science* 300 (12 June 2003): 1713. Sharp and Hahn report that two subtypes of one type of HIV-1 have a common ancestor, which implies that this virus was already in chimpanzees at some point early in the twentieth century. The authors argue that while human exposure to infected blood of the chimpanzees surely led to individual cases before the ones first recorded, it took demographic densities in the Leopoldville region until the mid-twentieth century to reach densities sufficient to kick-start a potential epidemic. See Sharp and Hahn, "Prehistory of HIV1," 605–606.

87. Moore, *Super Bugs*, 139.

Chapter 6

1. Paul J. Crutzen, "Geology of Mankind," *Nature* 415 (3 Jan. 2002): 23.

2. Jeremy Rifkin, *The Biotech Century: Harnessing the Gene and Remaking the World* (New York: Penguin Putnam, 1998), 73, 115.

3. See the tributes to this woman and her work in Peter Matthiessen (ed.), *Courage for the Earth* (Boston: Houghton Mifflin Company, 2007).

4. Rachel Carson, *Silent Spring*, fortieth anniversary edition (Boston: Houghton Mifflin Company, 2002), 68–69.

5. See Barry Commoner's fine early ecological work, *The Closing Circle: Nature, Man, and Technology* (New York: Bantam Books, 1974), 29–35.

6. In his book on chaos theory James Gleick writes that the butterfly effect was the starting point of converging thought for those various scientists who pioneered the "new science." James Gleick, *Chaos: Making a New Science* (New York: Penguin Books, 1987), 8.

7. For the following discussion, see "And No Birds Sing," chapter seven in Carson, *Silent Spring*, 103–118. It is interesting to note in this context that the title *Silent Spring* was not the first chosen by Carson and her publishers. Their initial preference was *Control of Nature*. Frank Graham, Jr., *Since Silent Spring* (Greenwich, Conn.: Fawcett Publications, 1970), 35.

8. Carson, *Silent Spring*, 106.

9. For the most comprehensive list of the hundreds of species of arthropods now resistant to pesticides, see the Michigan State database at http://www.pesticideresistance.org/security/1/.

10. Robert Platt and John Griffiths, *Environmental Measure and Interpretation* (New York: Reinhold Pub. Corp., 1964), 35.

11. "Honey Bee Colony Losses in U. S. Almost 30 Percent from All Causes from September 2008 to April 2009," *Science Daily*, 29 May 2009, 1 of 2, http://www.sciencedaily.com/releases/2009/5/09052218 0624.htm.

12. "Questions and Answers: Colony Collapse Disorder," *USDA Agricultural Research Service*, 20 Aug. 2009, 2 of 4, http://www.ars.usda.gov/News/docs.htm?docid=15572.

13. Ibid.

14. The following discussion of IMD draws from Michael Schacker, *A Spring without Bees: How Colony Collapse Disorder Has Endangered Our Food Supply* (Guilford, Conn.: The Lyons Press, 2008).

15. Ibid., 73–75.

16. Ibid., 78–79.

17. Ibid., 87.

18. Ibid., 132–134.

19. Quoted in "Colony Collapse Disorder: A Complex Buzz," *USDA Agricultural Research Service*, 7 July 2008, 1 of 3, http://www.ars.usda.gov/is/AR/archive/may/08/colony0508htm.

20. Quoted in Bill Graham, "Empty Hives Have No Easy Fix," *The Kansas City Star*, 7 July 2008, B6.

21. Schacker, *A Spring without Bees*, 134.

22. CCD Steering Committee, "Colony Collapse Disorder: Progress Report," 2 Dec. 2009, 4–10 of 22, http://www/extension.org/mediawiki/files/c/c7/CCDReport2009.pdp.

23. A good brief overview of the entire topic of CFCs and the ozone is in G. Tyler Miller, Jr., *Sustaining the Earth: An Integrated Approach*, 7th ed. Florence, KY: Thomson Brooks/Cole, 2005), 269–273.

24. Michael Weisskopf, "CFC's: Rise and Fall of Chemical 'Miracle'; Chlorofluorocarbons Vs. Ozone," 10 April 1980, 1 of 3, http://www.encyclopedia.com/doc.1P2–1250144.html.

25. F. Sherwood Rowland, "Stratospheric Ozone Depletion by Chlorofluorocarbons (Nobel Lecture)," 11 Aug. 2008, 1–2 of 17, httm://www.coearth.org/ article/stratosphere_Ozone_Depletion_by_Clorofluoracarbons_(Nob el _Lecture).

26. Ibid., 2 of 17.

27. Elizabeth Kolbert, *Field Notes from a Catastrophe* (New York: Bloomsbury, 2006), 184.

28. See Dr. Elmar Uherek, "Chlorofluorocarbons (CFC's) and the Ozone Hole," 20 Feb. 2006, 5 of 5, http://www.atmosphere.mpg. de/enid/lz2.htm/.

29. Ibid., 3–5 of 5.

30. See Richard Kerr's running commentary on the unfolding discussion of these matters as it took place within and around scientific findings. "Taking Shots at Ozone Hole Theories," *Science* 234 (14 Nov. 1986): 817–818; "Halocarbons Linked to Ozone Hole," *Science* 236 (5 June 1987): 1182–1183; and "Winds, Pollutants Drive Ozone Hole," *Science* 238 (9 Oct. 1987): 156–158.

31. Kolbert, *Field Notes from a Catastrophe*, 185.
32. "Ozone Depletion," *Wikipedia*, 10 Aug. 2008, 10 of 16, http://wikipedia.org/ Wiki/ Ozone_layer_depiction.
33. Miller, *Sustaining the Earth*, 272.
34. All figures here are from Paul Brown, "Ozone Depletion," *Conservation Science Institute*, 10 Aug. 2008, 2–3 of 3, http://www.conservation-institute. org/ocean_change/ocean_pollution. See also "Stratospheric Ozone Depletion," 12 Aug. 2008, 5 of 5, http://www.medicalecology.org/ Atmosphere/a_app_strat.htm.
35. Ibid.
36. http://news.bbc.co.uk/2/hi/science/ nature/5298004.stm Environmental science teacher Elizabeth McIntyre gave me this site.
37. "Stratospheric Ozone Depletion," 4 of 5.
38. As late as 1987, serious division split the Reagan administration concerning whether or not to go along with the aims of those about to meet in Montreal. See Mark Crawford, "Ozone Plan Splits Administration," *Science* 236 (29 May 1987): 1052–1053.
39. All figures here are from "Ozone Depletion," 8 of 16.
40. Rowland, "Stratospheric Ozone Depletion by Chlorofluorocarbons (Nobel Lecture)," 2 of 17.
41. Crutzen, "Geology of Mankind," 23.
42. The following discussion relies on this work. See Theo Colborn, Dianne Dumanoski, and John Peterson Myers, *Our Stolen Future: Are We Threatening Our Fertility, Intelligence, and Survival?— A Scientific Detective Story* (New York: Dutton, 1996). Although this collaborative work speaks of Colborn in the third person, she is the lead author for obvious reasons, and we will treat her as such when speaking about or quoting from the work.
43. See Gore's "Forward," ibid., v–vii.
44. Ibid., 15. Lest our own discussion here mislead, Colborn's deep appreciation of Rachel Carson and her work is quite evident in Colborn's acceptance of the Rachel Carson Leadership Award in 1997. See "Theo Colborn Receives the Rachel Carson Leadership Award," 19 June 2008, http://www.pmac.net/nctc.htm.
45. Colborn, *Our Stolen Future*, 202.
46. See ibid., 1–10.
47. See ibid., 11–28.
48. Ibid., 203–204.
49. Ibid., 204.
50. For a discussion of these two cases, see ibid., 47–67. Elizabeth McIntyre pointed out to me that thalidomide is a DNA-inhibitor, and that DES, an estrogen mimic, is, at least technically, not a hormone-disruptor.
51. Ibid., 51.
52. Ibid., 49.
53. Ibid., 66.
54. Ibid., 207.
55. Ibid., 122. See the extended discussion, 122–131.
56. Sarah Vogel, "Battles of Bisphenol A," 26 June 2008, www.defendingscience. Org/ case_studies_Battles_Over_Bisphenol_A.cfm.
57. Eartha Jane Melzer, "Momentum builds for tighter regulation of industrial Chemicals," *Michigan Messenger*, 14 Oct. 2009, 3 of 7, http:/ /michiganmessenger.com/27861/momentum-builds-for-tighter-regulation-of-industry.
58. Colborn, *Our Stolen Future*, 196, 197.
59. Ibid., 242, 243.
60. John McHale, *The Future of the Future* (New York: Ballantine Books, 1969), 5.
61. Ibid., 110.
62. Ervin Laszlo, *The Systems View of the World* (New York: George Brazillier, Inc., 1972), 120.

Chapter 7

1. On the founding of General System Theory and its fundamental ideas, see Ludwig von Bertalanffy, *General System Theory* (New York: Braziller, 1968), 14–17. The Society's "yearbook," *General Systems*, was founded two years later under the editorship of Bertalanffy and Rapoport. The Society refashioned itself as the International Society for Systems Science in 1988.
2. These new stonemasons saw the outline for their foundation laid by astrophysics, which, in Bertalanffy's phrasing, had progressively *deanthropomorphized* reality. "In its deanthropomorphized form, science is a conceptual construct representing certain formal or structural relationships in an unknown X," hence, "we are seeking for another basic outlook—*the world as organization.*" Ludwig von Bertalanffy, *Robots, Men and Minds* (New York: Braziller, 1967), 98, 57. A much fuller discussion of this attempt to refound scientific enlightenment can be found in my unpublished dissertation: Robert M. Tostevin, "Living without the Mastery of Nature: Beyond the Spell of Enlightenment," Dept. of Political Science, York University, Downsview, Ontario, 1977. The present study draws its themes and fundamental ideas from that work, which was primarily a critique of systems thinking and belief in mastery.
3. Bertalanffy, *General System Theory*, 48–49.
4. Anatol Rapoport, "The Search for Simplicity," in *The Relevance of General Systems Theory*, edited by Erwin Laszlo (New York: Braziller, 1972), 24.
5. Erwin Laszlo, *A Strategy for the Future* (New York: Braziller 1974), 17.
6. Bertalanffy, *General System Theory*, 33.
7. A. D. Hall and R. E. Fagen, "Definition

of System," in *General Systems*, vol. 1, edited by Ludwig von Bertalanffy and Anatol Rapoport (Washington, D.C.: Society for General Systems Research, 1956): 20.

8. Ludwig von Bertalanffy, *Perspectives on General System Theory: Scientific-Philosophical Studies* (New York: Braziller, 1975), 46.

9. Bertalanffy quoting himself from an earlier writing: *Robots*, 87–88.

10. Erich Jantsch, *Design for Evolution* (New York: Braziller, 1975), 34, 35.

11. Victor C. Ferkiss, *Technological Man: The Myth and the Reality* (New York: Mentor Books, 1969), 200, 201.

12. Ibid., 202, 203.

13. Ibid., 205–206, 207.

14. Bertalanffy, *Perspectives*, 164.

15. Laszlo, *A Strategy for the Future*, 60. The spaceship metaphor actually fuses together open system thinking (negentropic self-organization) with a closed system image (entropic spacecraft). This con-fusion seemed not to bother those employing it, probably because the image is so striking.

16. On the foundation and themes of the Santa Fe Institute, see M. Mitchel Waldrop, *Complexity: The Emerging Science at the Edge of Order and Chaos* (New York: Simon & Schuster, 1992). Complexity theory eschews prediction in the face of unpredictable phenomena and offers fascinating ideas attractive to those whose science stands "outside the box."

17. Paul J. Crutzen, "Geology of Mankind," *Nature* 415 (3 Jan. 2002): 23.

18. Thomas Dietz, Elinor Ostrom, and Paul C. Stern, "The Struggle to Govern the Commons," in *Science Magazine's State of the Planet: 2006–2007*, edited by Donald Kennedy (Washington: Island Press, 2006), 127.

19. Ibid., 137.

20. Anthony J. McMichael, Colin D. Butler, and Carl Folke, "New Visions for Addressing Sustainability," ibid., 163.

21. Ibid., 164.

22. Ibid., 165.

23. Norman Myers and Jennifer Kent (eds.), *The New Atlas of Planet Management* (Berkeley: University of California Press, 2005), 18.

24. Ibid., 12, 14.

25. Ibid., 282, 288.

26. Ibid., 289.

27. Ibid., 291.

28. Bill McKibben, *The End of Nature*, 2006 ed. (New York: Random House, 2006), 136.

29. Raymond Studer, "Human System Design and the Management of Change," in Bertalanffy and Rapoport, *General Systems*, vol. XVI, 141, 142. An overview of the origin and early development of systems analysis is in Charles Hitch, "An Appreciation of Systems Analysis," in *Systems Analysis*, edited by Stanford L. Optner (Baltimore: Penguin, 1973), 19–36. Two classic works on the nature and uses of systems technology in business organizations are: Stanford Optner, *Systems Analysis for Business Management* (Englewood Cliffs, NJ: Prentice-Hall, 1960); and Stafford Beer, *Management Science: The Business Use of Operations Research* (Garden City, NJ: Doubleday & Company Inc., 1969). For a more detailed treatment of the material discussed in this section, see Tostevin, "Living without the Mastery of Nature."

30. Steven L. Dickerson and Joseph E. Robertshaw, *Planning and Design* (Lexington: Lexington Books. 1975), 19.

31. Hall and Fagen, "Definition of System," 18. The authors immediately explain that their being "terse and vague" is unavoidable because the concept "simply is not amenable to complete and sharp definition."

32. "New Age" Bucky Fuller, whose idiosyncratic intelligence still attracts many followers, called on people to think big enough to eliminate any and all environmental variables that otherwise would escape control. "Spaceship Earth" arose out of just such thinking, which he undertook in the name of "a general system theory." See Buckminster Fuller, *Operating Manual for Spaceship Earth* (New York: Simon and Schuster, 1969).

33. David Orrell's work has been most helpful to the discussion here, as well as to a later discussion of modeling. See his *The Future of Everything: The Science of Prediction* (New York: Thunder's Mouth Press, 2007).

34. Herbert Maisel and Guiliano Gnugnoli, *Simulation of Discrete Stochastic Systems* (Chicago: Science Research Associates, 1972), 4.

35. G. Sommerhoff, "The Abstract Characteristics of Living Systems," in *Systems Thinking*, edited by F. E. Emery (Middlesex: Penguin, 1969), 175–176.

36. Dimitris Chorafas, *Systems and Simulation* (New York: Academic Press, 1965), 15.

37. Ibid., 8.

38. George Morganthaler, "The Theory and Application of Simulation in Operations Research," in *Progress in Operations Research*, vol. 1, edited by Russell Ackoff (New York: Wiley and Sons, 1961), 375.

39. Maisel and Gnugnoli, *Simulation of Discrete Stochastic Systems*, 33.

40. Wiener's seminal work appeared in 1948. Along with information theorists, Wiener identified what general system theorists call open systems with a closed machine system, the measure of information being conceived as a measure of negative entropy. See Norbert Wiener, *Cybernetics: Or Control and Communication in the Animal and the Machine* (Cambridge, MA: MIT Press, 1961), especially 1–29 and 169–180. More accessible is Wiener, *The Human Use of Human*

Beings: Cybernetics and Society (Garden City, NY: Doubleday & Company Inc., 1954), 15–47.

41. C. West Churchman, *The Systems Approach* (New York: Delta Books, 1968), 47.

42. Ibid., 174, 175.

43. See note 63 to chapter 4.

44. Mihajlo Mesarovich and Eduard Pestel, *Mankind at the Turning Point: The Second Report of the Club of Rome* (New York: E. P. Dutton & Co., Inc., 1974), 49.

45. Donella Meadows, Jorgen Randers, and Dennis Meadows, *Limits to Growth: The 30-Year Update* (White River Junction, VT: Chelsea Green Publishing, 2004) 48.

46. Ibid., 237.

47. Ibid., 269.

48. Ibid., 176.

49. See ibid., 259–263.

50. Ibid., 259–260.

51. Spencer R. Weart, *The Discovery of Global Warming* (Cambridge, MA: Harvard University Press, 2003) 168. For an overview of how global warming fared in Washington DC from the elder Bush, through Clinton, to the younger Bush, see David Howard Davis, *Ignoring the Apocalypse* (Westport, CT: Praeger, 2007) 187–201. A slashing treatment of the business behind much of the naysaying is Sharon Begley, "The Truth about Denial," *Newsweek* (13 Aug. 2007): 20–29.

52. See the general discussion in Thomas R. Karl and Kevin E. Trenberth, "Modern Global Climate Change," in *Science Magazine's State of the Planet: 2006–2007*, edited by Donald Kennedy (Washington, DC: Island Press, 2006), 88–98.

53. Some fully accept anthropogenic warming but argue against the negative balance sheet of the scientific consensus. See, for example, Bjorn Lomborg, *Cool It: The Skeptical Environmentalist's Guide to Global Warming* (New York: Alfred A. Knopf, 2007).

54. S. Fred Singer and Dennis T. Avery, *Unstoppable: Every 1,500 Years, Global Warming*, expanded ed. (New York: Rowman and Littlefield Publishers Inc., 2008), 253.

55. Weart, *The Discovery of Global Warming*, 199.

56. See ibid., 80–89.

57. Elizabeth Kolbert, *Field Notes from a Catastrophe* (New York: Bloomsbury, 2006), 50–51.

58. See Lawrence Solomon, *The Deniers* (Minneapolis, MN: Richard Vigilante Books, 2008), 161–175. Solomon, a long-time public activist, environmentalist, and leading figure in Energy Probe in Toronto, Ontario, gives various scientific naysayers more than a sympathetic platform in this book. Remaining uncommitted, Solomon himself believes that much scientific work needs still to be done before judgment should be made.

59. Ibid.; 64–74.

60. See these graphs in Kolbert, *Field Notes from a Catastrophe*, 29–131.

61. Weart, *The Discovery of Global Warming*, 130, 131.

62. Richard A. Kerr, "How Hot Will the Greenhouse World Be?" *Science* 305 (July 2005): 100.

63. John Abatzogluo, Joseph F. C. DiMento, Pamela Doughman, and Stefano Nespor, "A Primer on Global Climate Change and Its Likely Impacts," in *Climate Change: What It Means for Us, Our Children, and Our Grandchildren*, edited by Joseph F. C. DiMento and Pamela Doughman (Cambridge, MA: MIT Press, 2007), 39.

64. Weart, *The Discovery of Global Warming*, 191–192.

65. See Orrell, *The Future of Everything*, 134–137.

66. See the discussion in Kolbert, *Field Notes from a Catastrophe*, 101–105.

67. G. Tyler Miller, Jr., *Sustaining the Earth: An Integrated Approach*, 7th ed. Florence, KY: Thomson Brooks/Cole, 2005), 260–262.

68. Weart, *The Discovery of Global Warming*, 176.

69. See James Gleick, *Chaos: Making a New Science* (New York: Penguin Books, 1987), 11–23.

70. Richard A. Kerr, "Storm-in-a-Box Forecasting," *Science* 304 (May 2004): 948.

71. The Santa Fe Institute studies such phenomena. On cloud systems and weather forecasting in general, see Orrell, *The Future of Everything*, 149–159.

72. Ibid., 115.

73. Reiner Schnur writes, "The consensus approach to solving this problem [of modeling uncertainty] is to assume that the influence of modeling uncertainty can be limited by averaging several different models that differ slightly in their representation of the climate." Reiner Schnur, "The Investment Forecast," *Nature* 415 (31 Jan. 2002): 483.

74. Orrell, *The Future of Everything*, 159.

75. Ibid., 169.

76. Quoted ibid., 294.

77. Miller, *Sustaining the Earth*, 258.

78. Richard A. Kerr, "Scientists Tell Policymakers We're All Warming the World," *Science* 315 (9 Feb. 2007): 755.

79. See Solomon on the politics and science behind this 2007 deletion. Solomon, *The Deniers*, 29–36.

80. Richard A. Kerr, "Pushing the Scary Side of Global Warming," *Science* 316 (8 June 2007): 1412–1415.

81. Stefan Rahmstorf et al., "Recent Climate Observations Compared to Projections," *Science* 316 (4 May 2007): 709.

82. Naomi Oreskes, "The Scientific Consensus on Climate Change: How Do We Know

We're Not Wrong," in DiMento and Doughman, *Climate Change*, 79.

83. Ibid.

84. Joseph F. C. DiMento and Pamela Doughman. "Climate Change: What It Means for Us, Our Children, and Our Grandchildren," in DiMento and Doughman, *Climate Change*, 182.

85. Ibid., 183.

86. Oreskes, "The Scientific Consensus on Climate Change," 92, 93.

87. Solomon concludes that such disagreement is a problem for the naysayers. As he writes, "They can't all be right." Solomon, *The Deniers*, 208. One point does unite a number of the naysayers, though it is nothing in the way of an alternative explanation of global warming. This is their shared concern for the severe damage they see following any serious shift in the energy base of our economy. Belief in the so-called free market lies at the heart of such criticism. For two recent examples of such free-market thinking, see the meteorologist and climate scientist Roy W. Spencer, *Climate Confusion* (New York: Encounter Books, 2008); and Nigel Lawson, *An Appeal to Reason: A Cool Look at Global Warming* (New York: Overlook Duckworth, 2008). Both authors see most consensus thinking as religious commitment and fault it for, among other things, overlooking major economic matters. In this last connection it is reassuring to know that Lord Lawson's service to Margaret Thatcher's governments in various capacities prepared him to speak so convincingly on the real and pressing problem of poverty.

88. Andrew C. Revkin. "Climate Change as News: Challenges in Communicating Environmental Science" in DiMento and Doughman, *Climate Change*, 151.

89. Ibid., 157–158.

Conclusion

1. Hannah Arendt, *The Human Condition* (Chicago: University of Chicago Press, 1958), 207.

2. See ibid., 155–170.

3. Ibid., 207–208.

4. Ibid., 208.

5. Good coverage of such programmatic precautionary thinking is available in Timothy O'Riordan and James Cameron (eds.), *Interpreting the Precautionary Principle* (London: Earthscan Publications Ltd., 1994). Especially illuminating from our standpoint is the introductory essay written by the editors, as well as David Pearce's essay, which relates the precautionary principle to economics, and the piece by Daniel Bodansky, which addresses the principle in the context of environmental law in the United States.

6. See William Leiss, "Risk Issue Management," in his *In the Chamber of Risks: Understanding Risk Controversies* (Montreal and Kingston: McGill-Queens University Press, 2001), 3–15.

7. Langdon Winner, *The Whale and the Reactor: A Search for Limits in an Age of High Technology* (Chicago: University of Chicago Press, 1986), 148. See his larger discussion, 138–155.

8. See the survey provided in Alan Drengson and Yuichi Inoue (eds.), *The Deep Ecology Movement: An Introductory Anthology* (Berkeley, CA.: North Atlantic Books, 1995). The significance of "deep" as distinct from "shallow" ecology respects the level of philosophical reform such thinking strives for, as deep ecologists rest the possibility of serious ecological change on our fundamental categories of understanding. They quite typically address what the Western philosophical tradition discussed under ontology and metaphysics. Deep ecologists tend in this regard, however, to be largely appropriative in their thinking.

9. The classic work here is James Lovelock, *Gaia: A New Look at Life on Earth* (Oxford: Oxford University Press, 1979).

10. Stephen Jay Gould, "The Golden Rule: A Proper Scale for Our Environmental Crisis," in *Eight Little Piggies: Reflections in Natural History*, edited by Stephen Jay Gould (New York: W. W. Norton & Company, 1993), 48–49.

11. James Lovelock, *The Revenge of Gaia: Earth's Climate Crisis and the Fate of Humanity* (New York: Basic Books, 2006), 152.

12. David Orrell, *The Future of Everything: The Science of Prediction* (New York: Thunder's Mouth Press, 2007), 334, 348.

13. Bill McKibben, *The End of Nature*, 2006 ed. (New York: Random House, 2006), 50.

14. Ibid., xviii.

15. Ibid., 141.

16. Ibid., 144.

17. Ibid., 183.

18. Ibid., 182, 183.

19. Ibid., 164.

20. On the exercise front, Dr. Evans of the Howard Hughes Medical Institute and Salk Institute recently reported a promising experiment with AICAR, a new drug, carried out with mice. Aicar is "exercise in a pill;" it allows sedentary animals, or at least mice, to improve their physical performance and expand their endurance. "Assuming it becomes available for human use," says Dr. Richard N. Bergman, an expert on obesity and diabetes, "[i]t is possible that the couch potato segment of the population might find this to be a good regimen, and of course that is a large number of people." Nicholas Wade, "Couch Mouse to Mr. Mighty by Pills Alone,"

The New York Times, August 1, 2008, 3 of 5, http://www.nytimes.com/2008/08/01/science/01muscle.html?em.

21. Friedrich Nietzsche, *The Will to Power*, translated and edited by Walter Kaufmann (New York: Vintage, 1968), Aphorism, 84, 52.

22. Friedrich Nietzsche, *Beyond Good and Evil*, translated by Walter Kaufmann (New York: Vintage, 1966), Section 19, 26.

23. Arendt argues that, like thinking, willing entails a two-in-one split within ourselves, but that unlike the dialogical nature of the thinking faculty, in which we essentially talk to ourselves, the faculty of willing entails conflict. She writes, "The split occurs in the will itself; the conflict arises neither out of a conflict between mind and will nor out of a split between flesh and mind.... It is in the Will's nature to double itself, and in this sense, wherever there is a will there are always 'two wills'.... The trouble is that it is the same willing ego that simultaneously wills and nills.... [Hence, just as] it is in the nature of the will to command and demand obedience, it is also in the nature of the will to be resisted." Hannah Arendt, *Willing* (New York: Harcourt Brace Jovanovich, 1978), 94, 95–96.

24. Gide's story follows Michel's descent into the Dionysian, a descent that is no seduction of the will by the body but, rather, a willing descent that unfolds according to Michel's doctrinal will-to-strength. Having overcome all moral and traditional obstacles to such a will within himself, Michel's will negates itself in its pyrrhic victory. In the very end he appeals to friends: "Tear me away from this place now, give me some reason to live.... I can't leave of my own accord. Something in my will has been broken...." André Gide, *The Immoralist* (New York: Vintage International, 1996), 169, 170.

25. At one point the underground man tells us, "all my life it is the laws of nature that have offended me, continually, and more than anything else." What offends him is the necessity obtaining in such laws. Such necessity denies the will its efficacy. But the underground man, nevertheless, will have none of it. "[W]hat do I care about the laws of nature and about arithmetic when for some reason I don't like these laws or 'twice two is four'?" And he immediately responds, "I won't knock this wall down with my head if in the end I haven't got the strength to do so, but I won't submit to it simply because I'm up against a stone wall and haven't got sufficient strength." This perverse character's very perversity consists in his experiencing the pleasure of commanding defiance. Fyodor Dostoevsky, *Notes from the Underground and The Gambler*, translated by Jane Kentish (Oxford: Oxford University Press, 1991), 18, 15.

26. Bill McKibben, *Enough: Staying Human in An Engineered Age* (New York: Henry Holt and Company, 2003), 205.

Bibliography

Abatzogluo, John, et al. "A Primer on Global Climate Change and Its Likely Impacts." In *Climate Change: What It Means for Us, Our Children, and Our Grandchildren*, edited by Joseph F. C. DiMento and Pamela Doughman. Cambridge, Mass.: MIT Press, 2007.

Agassi, Joseph. *The Continuing Revolution: A History of Physics from the Greeks to Einstein*. New York: McGraw-Hill, 1968.

Alexander, H. G., ed. *The Leibniz-Clarke Correspondence*. New York: Manchester University Press, 1956.

Arendt, Hannah. *Between Past and Future*. New York: Viking Press, 1968.

_____. *The Human Condition*. Chicago: University of Chicago Press, 1958.

_____. *Willing*. New York: Harcourt Brace Jovanovich, 1978.

Ashton, T.S. *The Industrial Revolution, 1760–1830*. New York: Oxford University Press, 1964.

Avise, John C. *The Hope, Hype, and Reality of Genetic Engineering*. Oxford: Oxford University Press, 2004.

Bacon, Francis. *Essays and New Atlantis*. Roslyn, NY: Walter J. Black Pub., 1942.

_____. *The New Organon*. Indianapolis: The Bobbs-Merrill Company, Inc., 1960.

Bailes, Elizabeth, et al., "Hybrid Origin of HIV in Chimpanzees." *Science* 300 (12 June 2003).

Barry, John M. *The Great Influenza: The Story of the Deadliest Pandemic in History*. London: Penguin, 2005.

Baumer, Franklin L. *Modern European Thought: Continuity and Change in Ideas, 1600–1950*. New York: Macmillan, 1977.

Becker, Carl L. *The Heavenly City of the Eighteenth-Century Philosophers*. New Haven: Yale University Press, 1932.

Beer, Stafford. *Management Science: The Business Use of Operations Research*. Garden City, NY: Doubleday & Company Inc., 1967.

Begley, Sharon. "The Truth about Denial." *Newsweek* (August 13, 2007).

Bertalanffy, Ludwig von. *General System Theory*. New York: Braziller, 1968.

_____. *Perspectives on General System Theory: Scientific-Philosophical Studies*. New York: Braziller, 1975.

_____. *Robots, Men and Minds*. New York: Braziller, 1967.

Bishop, Morris. *The Middle Ages*. Boston: Houghton Mifflin, 1968.

Bohr, Niels. "Can Quantum-Mechanical Description of Physical Reality be Considered Complete?" *Physical Review* 48 (Oct 15, 1935).

_____. "Discussion with Einstein on Epistemological Problems in Atomic Physics." In *Albert Einstein: Philosopher-Scientist*, edited by P. A. Schilpp. Evanston: The Library of Living Philosophers Inc., 1949.

_____. *The Philosophical Writings of Niels Bohr: vol. I, Atomic Theory and the Description of Nature*. Connecticut: Ox Bow Press, 1987.

Bolles, Edmund Blair. *Einstein Defiant: Genius versus Genius in the Quantum Revolution*. Washington DC: Joseph Henry Press, 2004.

Bray, R. S. *Armies of Pestilence: The Impact of Disease on History*. New York: Barnes & Noble Books, 1996.

Brinton, Crane. *The Shaping of Modern Thought*. Englewood Cliffs: Prentice-Hall, 1963.

Bronowski, Jacob. *The Common Sense of Science*. New York: Vintage Books, 1950.

Brown, Paul. "Ozone Depletion." *Conserva-*

tion Science Institute. 10 Aug. 2008 http://www.conservationinstitute.org/ocean_change/ocean_pollution.
Burchell, S. C. *Age of Progress*. New York: Time Incorporated, 1966.
Burke, Edmund. *Reflections on the French Revolution in France*. USA: Bobbs-Merrill Co., Inc., 1955.
Burtt, E. A. *The Metaphysical Foundations of Modern Science*: Garden City, NY: Doubleday and Co., 1954.
Bury, J. B. *The Idea of Progress*. New York: Dover Books, 1932.
Butler, Declan. "Flu Officials Pull Back from Raising Global Alert." *Nature* 436 (7 July 2005).
_____. "Yes, but Will It Jump?" *Nature* 439 (12 Jan. 2006).
Butterfield, Herbert. *The Origins of Modern Science: 1300–1800*. Rev. ed. Toronto: Clarke, Irwin, & Co., 1968.
Čapek, Milič. *The Philosophical Impact of Contemporary Physics*. New York: Van Nostrand Reinhold, 1961.
Carson, Rachel. *Silent Spring*. Fortieth Anniversary Edition, Boston: Houghton Mifflin Company, 2002.
Cassirer, Ernst. *The Philosophy of the Enlightenment*, translated by Fritz Koelin and James Pettegrove. Boston: Beacon Press, 1951.
Champion, Justin A. I. "Deism." In *The Columbia History of Western Philosophy*, edited by Richard Popkin. New York: Columbia University Press, 1999.
Chorafas, Dimitris. *Systems and Simulation*. New York: Academic Press, 1965.
Christianson, Gale E. *In the Presence of the Creator: Isaac Newton and His Times*. London: Collier Macmillan, 1984.
Churchman, C. West. *The Systems Approach*. New York: Delta Books, 1968.
Cohen, H. Floris. *The Scientific Revolution: A Historiographical Enquiry*. Chicago: University of Chicago Press, 1994.
Cohen, I. Bernard. *The Birth of a New Physics*. Garden City, NY: Doubleday & Co., 1960.
_____. *Revolution in Science*. Cambridge, MA: Harvard University Press, 1985.
Colborn, Theo, Dianne Dumanoski, and John Peterson Myers. *Our Stolen Future: Are We Threatening Our Fertility, Intelligence, and Survival?— A Scientific Detective Story*. New York: Dutton, 1996.
"Colony Collapse Disorder: A Complex Buzz." *USDA Agricultural Research Service*. 7 July 2008, http://www.ars.usda.gov/is/AR/archive/may08/colony0508.htm.
"Colony Collapse Disorder: Progress Report." CCD Steering Committee. June 2009, http://www.extension.org/mediawiki/files/C/C7/CCDReport 2009.pdf.
"Colony Collapse Disorder: Why Are Honey Bees Disappearing?" 7 July 2008 http://solutionsforourlife.ufl.edu/hot_topics/agriculture/colony_collapse_ Disorder.html.
Commoner, Barry. *The Closing Circle: Nature, Man and Technology*. New York: Bantam Books, 1971.
_____. "In Defense of Biology." *Science* 133 (2 June 1961).
Condorcet, Marquis de. "The Progress of the Human Mind." In *Main Currents in Western Thought*, edited by Franklin Le Van Baumer, New Haven: Yale University Press, 1978.
Copernicus, Nicolas. "The *Commentariolus*." In *Three Copernican Treatises*, translated by Edward Rosen, New York: Dover Books, 1959.
"Cotes' Preface to the Second Edition of the *Principia*." In Isaac Newton, *Principia*, vol. I, translated by Andrew Motte, revised by Florian Cajori. Berkeley: University of California Press, 1962.
Crawford, Mark. "Ozone Plan Splits Administration." *Science* 236 (29 May 1987).
Crutzen, Paul J. "Geology of Mankind." *Nature* 415 (3 January 2002).
Cziko, Gary. *Without Miracles: Universal Selection Theory and the Second Darwinian Revolution*. Cambridge, MA.: MIT Press, 1995.
D'Alembert, Jean le Rond. "The Human Mind Emerged from Barbarism." In *The Portable Enlightenment Reader*, edited by Isaac Kramnick, New York: Penguin Books, 1995.
Dantas, Gautam, et al. "Bacteria Subsisting on Antibiotics." *Science* 230 (4 April 2008).
Darnton, Robert. *The Kiss of Lamourette: Reflections in Cultural History*. New York: W. W. Norton, 1990.
Darrow, Karl. "The Quantum Theory." *Scientific American Reprint* 205 (March 1952).
Darwin, Charles. *On the Origin of Species*. Cambridge, MA: Harvard University Press, 1964.
Davies, Brian. *Science in the Looking Glass:*

What Do Scientists Really Know? Oxford: Oxford University Press, 2003.

Davis, David Howard. *Ignoring the Apocalypse*. Westport: Praeger, 2007.

Deane, Phyllis. *The First Industrial Revolution*. London: Cambridge University Press, 1967.

Dear, Peter. *Revolutionizing the Sciences*. Princeton: Princeton University Press, 2001.

De Broglie, Louis. *The Revolution in Physics*, translated by Ralph Niemeyer. New York: Noonday Press, 1953.

Descartes, René. *Philosophical Essays*, translated by Laurence J. Lafleur. Indianapolis: Bobbs-Merrill Co., 1964.

Dickerson, Steven L., and Robertshaw, Joseph E. *Planning and Design*. Lexington: Lexington Books, 1975.

Deutsch, David. *The Fabric of Reality: The Science of Parallel Universes and Its Implications*. New York: Penguin, 1997.

Diderot, Denis. "*Encyclopedia*." In *The Portable Enlightenment Reader*, edited by Isaac Kramnick. New York: Penguin Books, 1995.

Dietz, Thomas, Elinor Ostrom, and Paul C. Stern. "The Struggle to Govern the Commons." *Science Magazine's State of the Future: 2006–2007*, edited by Donald Kennedy, Washington, DC: Island Press, 2006.

Dimento, Joseph F. C., and Pamela Doughman. "Climate Change: What It Means for Us, Our Children, and Our Grandchildren." In *Climate Change: What It Means for Us, Our Children, and Our Grandchildren*, edited by Joseph F. C. DiMento and Pamela Doughman. Cambridge, MA: MIT Press, 2007.

Dijksterhuis, E. J. *The Mechanization of the World Picture*, translated by C. Dikshoorn. London, Oxford University Press, 1964.

Dobbs, Betty Jo Teeter, and Margaret C. Jacob. *Newton and the Culture of Newtonianism*. New York: Humanities Books, 1995.

Donne, John. "An Anatomy of the Universe: First Anniversary." In *John Donne: The Complete English Poetry*, edited by A. J. Smith, London: Penguin Books, 1971.

Dostoevsky, Fyodor. *Notes from the Underground and The Gambler*, translated by Jane Kentish. Oxford: Oxford University Press, 1991.

Drabble, Margaret. "The Great Exposition." In *Victorian England*, edited by Clarice Sinisher. San Diego: Greenhaven Press, Inc., 2000.

Drengson, Alan, and Yuichi Inoue, eds. *The Deep Ecology Movement: An Introductory Anthology*. Berkeley: North Atlantic Books, 1995.

Duncan, David Ewing. *The Geneticist Who Played Hoops with My DNA*. New York: Harper Collins, 2005.

Einstein, Albert. "Autobiographical Notes." In *Albert Einstein: Philosopher-Scientist*, edited by P. A. Schilpp. Evanston: The Library of Living Philosophers, Inc., 1949.

_____. "Can Quantum-Mechanical Description of Nature be Considered Complete?" *Physical Review* 48 (Oct. 15, 1935).

_____. "On the Generalized Theory." *Scientific American Reprint* 209 (April 1950).

_____. *Relativity: The Special and the General Theory*. Tanslated by Robert W. Lawson. London: Methuen & Co. Ltd., 1954

_____, and Leopold Infeld. *The Evolution of Physics*. New York: Simon and Schuster, 1960.

Eiseley, Loren. *Darwin's Century: Evolution and the Men Who Discovered it*. Garden City, NY: Anchor Books, 1961.

_____. *The Man Who Saw Through Time*. New York: Scribner's Sons, 1973.

Eldredge, Niles. *The Pattern of Evolution*. New York: W. H. Freeman and Company, 2000.

Feigle, Herbert, and Grover Maxwell, eds. *Minnesota Studies in the Philosophy of Science: Scientific Explanation, Space, and Time*, vol. III. Minneapolis: University of Minnesota Press, 1962.

Feingold, Mordecai. *The Newtonian Moment: Isaac Newton and the Making of Modern Culture*. New York: Oxford University Press, 1984.

Feng Gao, et al. "Origin of HIV-1 in the Chimpanzee Pan troglodytes Troglodytes." *Nature* 397 (4 Feb. 1999).

Ferkiss, Victor C. *Technological Man: The Myth and the Reality*. New York: Mentor Books, 1969.

Ferry, David. *Gilgamesh: A New Rendering in English Verse*. New York: Farrar, Straus and Giroux, 1992.

Feyerabend, Paul K. "On a Recent Critique of Complementarity: Part II." *Philosophy of Science* XXXVI, no. 1 (March 1969).

Ficino, Marsilio. "The Soul of Man." In *The Portable Renaissance Reader*, edited by

James Bruce Ross and Mary Martin McLaughlin. New York: Penguin Books, 1968.

Fisher, George W. "Sustainable Living: Common Ground for Geology and Theology." In *The Earth Around Us: Maintaining a Livable Planet*, edited by Jill S. Schneiderman. New York: W. H. Freeman and Company, 2000.

Fritzsche, Peter. *Stranded in the Present: Modern Time and the Melancholy of History*. Cambridge, MA: Harvard University Press, 2004.

Fontenelle de, Bernard. "Digressions on the Ancients and Moderns." In *Main Currents in Western Thought*, edited by Franklin Le Van Baumer. New Haven: Yale University Press, 1978.

Fossell, Michael B. *Cells, Aging, and Human Disease*. Oxford: Oxford University Press, 2004.

Fuller, Buckminster. *Operating Manual for Spaceship Earth*. New York: Simon & Schuster, 1969.

Fuller, Steve. *Kuhn vs. Popper: The Struggle for the Soul of Science*. New York: Columbia University Press, 2004.

Galilei, Galileo. "The Assayer." In *Discoveries and Opinions of Galileo*, translated by Stillman Drake. New York: Doubleday and Co., 1957.

Garreau, Joel. *Radical Evolution*. New York: Doubleday, 2005.

Garrett, Laurie. *The Coming Plague: Newly Emergent Diseases in a World Out of Balance*. New York: Penguin, 1994.

Gascoigne, John. "A Reappraisal of the Role of Universities in the Scientific Revolution." In *Reappraisals of the Scientific Revolution*, edited by David C. Lindberg and Robert S. Westfall, 207–260. Cambridge: Cambridge University Press, 1990.

Gay, Peter. *The Enlightenment: An Interpretation*, vol. I, New York: Alfred Knopf Pub., 1966; and vol. II, 1969.

Gide, André. *The Immoralist*. New York: Vintage Internationalist, 1996.

Gillispie, Charles Coulston. *The Edge of Objectivity*. Princeton: Princeton University Press, 1960.

Gleick, James. *Chaos: Making a New Science*. New York: Penguin Books, 1987.

"The Global MDR-TB and XDR-TB Response Plan." *World Health*. 2007. http://www,wgi,ubt.tb.featyres_archive/-Global_response_plan/en/index.html.

"Goals of Science." *The International Encyclopedia of the Social Sciences*. 12 March 2005, http://en.wikipedia.org/wiki/science.

Gordon, Jon W. "Genetic Enhancement in Humans." *Science* 284 (26 March 1999).

Goudge, T. A. *The Ascent of Life*. Toronto: University of Toronto Press, 1961.

Gould, Stephen Jay. *Eight Little Piggies: Reflections in Natural History*. New York: W. W. Norton & Co., 1993.

_____. *This Wonderful Life: The Burgess Shale and the Nature of History*. New York: W. W. Norton & Co., 1989.

Graham, Bill. "Empty Hives Have No Easy Fix." *The Kansas City Star*. 7 July 2008. B6.

Graham, Frank, Jr. *Since Silent Spring*. Greenwich, Conn: Fawcett, 1970.

Green, Ronald S. *Babies by Design: The Ethics of Genetic Choice*. New Haven: Yale University Press, 2007.

Greene, John C. *The Death of Adam*. Ames, IA: Iowa State University Press, 1959.

Greene, Lisa. "SARS: A Learning Experience, a Lurking Fear." *St. Petersburg Times*. January 31, 2005. http://www.sptimes.com/2005/01/31/Worldandnation/SARS-A_learning_expe.shtml.

Gribben, John. *The Scientists*. New York: Random House, 2002.

Guarante, Leonard, and Cynthia Kenyon. "Genetic Pathways that Regulate Ageing in Model Organisms." *Nature* 408 (November 9, 2000).

Hackman, W. D. "Scientific Instruments: Models of Brass and Aids to Discovery." In *The Uses of Experimentation: Studies in the Natural Sciences*, edited by David Gooding et al. Cambridge: Cambridge University Press, 1989.

Haldane, Elizabeth S., and Frances H. Simson, eds. and trans. *Hegel's Lectures on the History of Philosophy*, vol. II. New York: Humanities Press, 1955.

Haldane, Elizabeth S., and G. R. T. Ross, trans. *Philosophical Works of Descartes*, vol. I. Mineola, NY: Dover Publications, 1955.

Hall, A. Rupert. *From Galileo to Newton*. New York: Dover Publications, Inc., 1981.

Hall, A. D. and R. E. Fagen. "Definition of System." In *General Systems. V. 1(1956)*, edited by Anatol Rapoport, Ludwig von Bertalanffy. Washington, D.C.: Society for General Systems Research, 1956.

Hankins, Thomas. *Science and the Enlight-*

enment. Cambridge: Cambridge University Press, 1985.
Hanson, Norwood. *Patterns of Discovery*. London: Cambridge University Press, 1965.
_____. "The Copenhagen Interpretation of Quantum Theory." In *Philosophy of Science*, edited by Arthur Danto and Sidney Morgenbesser. Cleveland: Meridian Books, 1960.
Hazard, Paul. *The European Mind: 1680–1715*, translated by J. Lewis May. Middlesex: Penguin Books, 1964.
Heisenberg, Werner. *Physics and Beyond*, translated by Arnold Pomerans. New York: Harper & Row, 1971.
_____. *Physics and Philosophy: The Revolution in Modern Physics*. New York: Harper & Row, 1958.
Heller, Erich. *The Artist's Journey into the Interior and Other Essays*. New York: Random House, 1965.
Hempel, Carl, "A Logical Critical Appraisal of Operationism." In *Readings in the Philosophy of Science*, edited by Baruch A. Brody. Englewood Cliffs, NJ: Prentice-Hall, 1970.
_____, and Paul Oppenheim. "Studies in the Logic of Explanation." In *Readings in the Philosophy of Science*, edited by Baruch A. Brody. Englewood Cliffs, NJ: Prentice-Hall, 1970.
Henry, John. *The Scientific Revolution and the Origins of Modern Science*. New York: Palgrave Press, 2002.
Herivel, John. *The Background to Newton's Principia*. London: Oxford University Press, 1965.
Hesse, Mary. *Forces and Fields: A Study of Action at a Distance in the History of Physics*. Totawa: Littlefield, Adams & Co., 1965.
Hitch, Charles. "An Appreciation of Systems Analysis." In *Systems Analysis*, edited by Stanford L. Optner. Baltimore: Penguin, 1973.
Hobsbawm, Eric. *The Age of Revolution: 1789–1848*. New York: Mentor, 1962.
Hooker, Richard, trans. *Jean La Rond d'Alembert: Preliminary Discourse to the Encyclopedia of Diderot*. 8 June 2005. http://www.wsu.edu8080/-wldciv/world_civ_reader/world_civ_reader.
Horgan, John. *The End of Science: Facing the Limits of Knowledge in the Twilight of the Scientific Age*. New York: Broadway Books, 1997.
Horkheimer, Max. *Eclipse of Reason*. New York: Seabury Press, 1974.
_____, and Theodor W. Adorno. *Dialectic of Enlightenment*, translated by John Cumming. New York: Herder and Herder, 1969.
"How Common Are Drug-Resistant Diseases?" *National Public Radio*. September 2007. http://www.npr.org/templates/story/story/php?storyId=10613348.
Hume, David. "On Miracles." In *Enquiries Concerning the Human Understanding and Concerning the Principles of Morals*, edited by L. A. Selby-Bigge. Oxford: Oxford University Press, 1957.
Jammer, Max. *Concepts of Space: The History of Theories of Space in Physics*. Cambridge, MA: Harvard University Press, 1969.
Jantsch, Erich. *Design for Evolution: Self-Organization and Planning in the Life of Human Systems*. New York: Braziller, 1975.
Jonas, Hans. *Philosophical Essays: From Ancient Creed to Technological Man*. Chicago: University of Chicago Press, 1974.
Kaku, Michel. *How Science Will Revolutionize the 21st Century*. New York: Anchor Books, 1997.
Kanahu, Annabel, and Jenni Fredriksson. "History of AIDS up to 1986." 6 May 2008. http://www.avert.org/aids-history-86.htm.
Kant, Immanuel. "What Is Enlightenment?" In *Foundations of the Metaphysics of Morals and What Is Enlightenment*, edited and translated by Lewis White Beck. Chicago: University of Chicago Press, 1950.
Karl, Thomas R., and Trenberth, Kevin E. "Modern Global Climate Change." In *Science Magazine's State of the Planet: 2006–2007*, edited by Donald Kennedy. Washington, DC: Island Press, 2006.
Keller, Helen Fox. "Nature, Nurture, and the Human Genome Project." In *Scientific and Social Issues in the Human Genome Project*, edited by Daniel J. Kevles and Leroy Hood. Cambridge, MA: Harvard University Press, 1992.
Kerr, Richard. "Halocarbons Linked to Ozone Hole." *Science* 236 (5 June 1987).
_____. "How Hot Will the Greenhouse World Be?" *Science* 309 (1 July 2005).
_____. "Pushing the Scary Side of Global Warming." *Science* 316 (8 June 2007).
_____. "Scientists Tell Policymakers We're All Warming the World." *Science* 315 (9 February 2007).

_____. "Storm-in-a-Box Forecasting." *Science* 304 (14 May 2004).
_____. "Taking Shots at Ozone Hole Theories." *Science* 234 (14 November 1986).
_____. "Winds, Pollutants Drive Ozone Hole." *Science* 238 (9 October 1987).
Kline, Morris. *Mathematics in Western Culture.* New York: Oxford University Press, 1953.
Knight, David. *The Age of Science.* Oxford: Basil Blackwell Ltd., 1986.
Knox, T. M., trans. *Hegel's Philosophy of Right.* London: Oxford University Press, 1964.
Kolata, Gina. *Flu.* New York: Simon & Schuster, 2001.
Kolbert, Elizabeth. *Field Notes from a Catastrophe.* New York: Bloomsbury, 2006.
Koyré, Alexander. *From the Closed World to the Infinite Universe.* Baltimore: John Hopkins Press, 1957.
_____. *Metaphysics and Measurement: Essays in the Scientific Revolution.* Cambridge, MA: Harvard University Press, 1969.
_____. *Newtonian Studies.* Chicago: University of Chicago Press, 1965.
Kuhn, Thomas. *The Copernican Revolution.* Cambridge, MA: Harvard University Press, 1957.
_____. *The Essential Tradition: Selected Studies in Scientific Tradition and Change.* Chicago: University of Chicago Press, 1977.
_____. *The Road Since Structure.* Chicago: University of Chicago Press, 2000.
_____. *The Structure of Scientific Revolutions.* 2nd ed. Chicago: University of Chicago Press, 1970.
Lakatos, Imre, and Alan Musgrave, eds. *Criticism and the Growth of Knowledge.* London: Cambridge University Press, 1975.
Laszlo. Erwin. *A Strategy for the Future.* New York: George Braziller, Inc., 1974.
_____. *The Systems View of the World.* New York: George Braziller, Inc., 1972.
Lawson, Nigel. *An Appeal to Reason: A Cool Look at Global Warming.* New York: Overlook Duckworth, 2008.
Leiss, William. *Domination of Nature.* Boston: Beacon Press, 1974
_____. *In the Chamber of Risks: Understanding Risk Controversies.* Montreal and Kingston: McGill-Queens University Press, 2001.
Levy, Elinor, and Fischetti, Mark. *The New Killer Diseases.* New York: Crown Pub., 2003.
Levy, Stuart B. *The Antibiotic Paradox: How the Misuse of Antibiotics Destroys Their Curative Powers.* 2nd ed. New York: Perseus Publications, 2002.
Lewontin, Richard. *Biology as Ideology.* New York: Harper Collins Pub., 1991.
_____. *It Ain't Necessarily So: The Dream of the Human Genome and Other Illusions.* New York: The New York Review of Books, 2001.
Lindberg, David C. "Conceptions of the Scientific Revolution from Bacon to Butterfield." In *Reappraisals of the Scientific Revolution*, edited by David C. Lindberg and Robert S. Westman, 1–26. Cambridge: Cambridge University Press, 1990.
_____. *Uncertainty: Einstein, Heisenberg, Bohr, and the Struggle for the Soul of Science.* New York: Doubleday, 2007.
Lomborg, Bjorn. *Cool It: The Skeptical Environmentalist's Guide to Global Warming.* New York: Alfred A. Knopf, 2007.
Lovelock, James. *Gaia: A New Look at Life on Earth.* Oxford: Oxford University Press, 1979.
_____. *The Revenge of Gaia: Earth's Climate Crisis and the Fate of Humanity.* New York: Basic Books, 2006.
Löwith, Karl. *Meaning in History.* University of Chicago Press, 1949.
Lukacs, John. *The End of the Twentieth Century.* New York: Ticknor & Fields, 1993.
Maisel, Herbert, and Guiliano Gnugnoli. *Simulation of Discrete Stochastic Systems.* Chicago: Science Research Associates, 1972.
Malherbe, Michel. "Bacon's Method of Science." In *The Cambridge Companion to Bacon*, edited by Markku Peltonen. Cambridge: Cambridge University Press, 1996.
Marx, Karl, and Friedrich Engels. "Manifesto of the Communist Party." In *Selected Works.* Moscow: Progress Publishers, 1968.
_____ and _____. "Theses on *Feuerbach.*" In *Selected Works.* Moscow: Progress Publishers, 1968.
Mathison, Peter, ed. *Courage for the Earth.* Boston: Houghton Mifflin Company, 2007.
Mayr, Ernst. "Methodological Issues in Biology." In *Man and Nature: Philosophical Issues in Biology*, edited by Ronald Munson. New York: Delta Books, 1971.
_____. *What Evolution Is.* New York: Basic Books, 2001.

McHale, John. *The Future of the Future*. New York: Ballantine Books, 1969.

McKibben, Bill. *The End of Nature*. New York: Random House, 2006.

_____. *Enough: Staying Human in An Engineered Age*. New York: Henry Holt and Company: 2003.

McMichael, Anthony J., Colin D. Butler, and Carl Folke. "New Visions for Science." In *Science Magazine's State of the Planet: 2006–2007*, edited by Donald Kennedy. Washington: Island Press, 2006.

Meadows, Donella, Jorgen Randers, and Dennis Meadows. *Limits to Growth: The 30-Year Update*. White River Junction, VT: Chelsea Green Publishing, 2004.

Melzer, Eartha Jane. "Momentum Builds for Tighter Regulation of Industrial Chemicals." *Michigan Messenger*. Oct. 14, 2009. http://michigan messenger.com/27861/momentum-builds-for-tighter-regulation-of-industry.

Mesarovich, Mihajlo, and Eduard Pestel. *Mankind at the Turning Point: The Second Report of the Club of Rome*. New York: E. P. Dutton and Co., Inc., 1974.

"Microbes: What Doesn't Kill Them Makes Them Stronger." NISE project funded by the National Science Foundation. 2007. http://whyfiles.org/038badbugs/-Mechanism.html.

Miller, G. Tyler, Jr. *Sustaining the Earth: An Integral Approach*. Florence, KY: Thomson Brooks/Cole, 2005.

Millhauser, Milton. "In the Air." In *Darwin*, edited by Philip Appleman. New York: W. W. Norton & Co., 1970.

Montaigne, Michel de. *The Complete Essays*, translated by M. A. Screech. London: Penguin Books, 1991.

Montesquieu, Baron de. *The Spirit of the Laws*. Cambridge: Cambridge University Press, 1989.

Moore, Pete. *Super Bugs: Rogue Diseases of the Twenty-first Century*. London: Carlton Books, 2001.

Morganthaler, George. "The Theory and Application of Simulation in Operations Research." In *Progress in Operations Research*, vol. 1, edited by Russell Ackoff. New York: Wiley and Sons, 1961.

Mumford, Lewis. *The Myth of the Machine: Technics and Human Development*. New York: Harcourt, Brace & World, Inc., 1967.

_____. *The Myth of the Machine: The Pentagon of Power*. New York: Harcourt, Brace & World, Inc., 1970.

_____. *Technics and Civilization*. NY: Harcourt, Brace & World, Inc., 1963.

Myers, Norman, and Jennifer Kent, eds. *The New Atlas of Planet Management*. Berkeley: University of California Press, 2005.

Naam, Ramez. *More Than Human: Embracing the Promise of Biological Enhancement*. New York: Broadway Books, 2005.

Nef, John U. *Western Civilization Since the Renaissance: Peace, War, Industry and the Arts*. New York: Harper and Row, 1963.

Newton, Isaac. "From a Letter to Oldenburg." In *Newton's Philosophy of Nature: Selections from His Writings*, edited by H. S. Thayer. New York: Hafner Press, 1953.

Newton, Isaac. "*Principia*." In *On the Shoulders of Giants: The Great Works of Physics and Astronomy*, edited by Stephen Hawking, 733–1160. Philadelphia: Running Press, 2002.

Nietzsche, Friedrich. *Beyond Good and Evil*, translated by Walter Kaufmann. New York: Vintage, 1966.

_____. *The Birth of Tragedy and The Case against Wagner*, translated by Walter Kaufmann. Vintage, 1967.

_____. *The Will to Power*, translated and edited by Walter Kaufmann. New York: Vintage, 1968.

Offit, Paul. *Vaccinated: One Man's Quest to Defeat the World's Deadliest Diseases*. New York: Harper Collins, 2007.

Oldstone, Michael B. *Viruses, Plagues, and History*. Oxford: Oxford University Press, 1998.

Oppenheimer, J. Robert. *Science and the Common Understanding*. New York: Simon Schuster, 1954.

Optner, Stanford L. *Systems Analysis for Business Management*. Englewood Cliffs, NJ: Prentice-Hall, 1960.

Oreskes, Naomi. "The Scientific Consensus on Climate Change: How Do We Know We're Not Wrong?" In *Climate Change: What It Means for Us, Our Children, and Our Grandchildren*, edited by Joseph F. C. DiMento and Pamela Doughman, . Cambridge, MA: MIT Press, 2007.

O'Riordan, Timothy, and James Cameron, eds. *Interpreting the Precautionary Principle*. London: Earthscan Publications, Ltd., 1994.

Orrell, David. *The Future of Everything: The Science of Prediction*. New York: Thunder's Mouth Press, 2007.

"Ozone Depletion." *Wikipedia*. http://wikipedia.org/wiki/Ozone_layer_depletion.

Pascal, Blaise. *Pensées*, translated by A. J. Krailsheimer. Middlesex: Penguin Books, 1966.

Platt, Robert, and John Griffiths. *Environmental Measure and Interpretation*. New York: Reinhold Pub. Corp., 1964.

Pérez-Ramos, Antonio. "Bacon's Forms and the Maker's Knowledge Tradition." In *The Cambridge Companion to Bacon*. Cambridge: Cambridge University Press, 2001.

Pope, Alexander. "Intended for Sir Isaac Newton." In *Alexander Pope: Selected Works*, edited by Louis Kronenberger. New York: Modern Library, 1951.

Popkin, Richard H. *The History of Skepticism from Erasmus to Spinoza*. Berkeley, CA: University of California Press, 1979.

Popper, Karl R. *Conjectures and Refutations: The Growth of Scientific Knowledge*. New York: Harper & Row, 1965.

_____. *The Logic of Scientific Discovery*. Toronto: University of Toronto Press, 1959.

Priestley, Joseph. "Essay on Government." In *Main Currents in Western Thought*, edited by Franklin Le Van Baumer. New Haven: Yale University Press, 1978.

"Questions and Answers: Colony Collapse Disorder." *USDA Agricultural Research Service*. 10 March 2010. http://www.ars.usda.gov/News/docs.htm?docid=15572.

Rabb, Theodore K. *The Struggle for Stability in Early Modern Europe*. New York: Oxford University Press, 1975.

Rahmstorf, Stefan, et al. "Recent Climate Observations Compared to Projections." *Science* 316 (4 May 2007).

Rapoport, Anatol. "The Search for Simplicity." In *The Relevance of General System Theory*, edited by Erwin Laszlo. New York: Braziller, 1972.

Revkin, Andrew C. "Climate Change as News: Challenges in Communicating Environmental Science." In *Climate Change: What It Means for Us, Our Children, and Our Grandchildren*, edited by Joseph F. C. DiMento and Pamela Doughman. Cambridge, MA: MIT Press, 2007.

Rifkin, Jeremy. *The Biotech Century*. New York: Penguin Putnam, 1998.

Rossi, Paolo. *Francis Bacon: From Magic to Science*, translated by Sasha Rabinovitch. Chicago: University of Chicago Press, 1968.

_____. *Philosophy, Technology and the Arts in the Early Modern Era*. New York: Harper, 1970.

Rowland, F. Sherwood. "Stratospheric Ozone Depletion by Chlorofluorocarbons (Nobel Lecture)." *Encyclopedia of Earth*. 7 August 2008. http://www.eoearth.org/article/Stratospheric_Ozone_Depletion_by_Chlorofluorocarbons.

Sachs, Mendel. *Einstein versus Bohr*. La Salle: Open Court Publishing Co., 1988.

Schacker, Michael. *A Spring without Bees: How Colony Collapse Disorder Has Endangered Our Food Supply*. Guilford: The Lyons Press, 2008.

Schama, Simon. *A History of Britain*, vol. III, New York: Hyperion, 2002.

Schnur, Reiner. "The Investment Forecast." *Nature* 415 (31 January 2002).

"Science." In *The International Encyclopedia of the Social Sciences*. New York: Crowell, Collier and Macmillan, 1968.

Scriven, Michael. "Explanation and Prediction in Evolutionary Theory." In *Man and Nature*, edited by Ronald Munson. New York: Delta Books, 1971.

Segrè, Gino. *Faust in Copenhagen: A Struggle for the Soul of Physics*. New York: Penguin Books, 2007.

"Severe Acute Respiratory Syndrome (SARS)." *CDC Factsheet*. 25 Sept. 2007. http://www.cec.gov/ncidod-/sars/faq.htm.

Shapere, Dudley. *Galileo: A Philosophical Study*. Chicago: University of Chicago Press, 1974.

_____. "Introduction." In *Philosophical Problems of Natural Science*, edited by Dudley Shapere. London: Macmillan, 1965.

Sharp, Paul M., and Beatrice H. Hahn. "Prehistory of HIV 1." *Nature* 455 (2 Oct. 2008).

Shapin, Steven. *The Scientific Life: A Moral History of a Late Modern Vocation*. Chicago: University of Chicago Press, 2009.

_____. *The Scientific Revolution*. Chicago: University of Chicago Press, 1996.

Silver, Lee M. *Challenging Nature: The Clash between Biotechnology and Spirituality*. New York: Harper Collins, 2006.

_____. *Remaking Eden: How Genetic Engineering and Cloning Will Alter the American Family*. New York: Harper Collins, 1998.

Simpson, George Gaylord. *The Major Features of Evolution*. New York: Simon & Schuster, 1953.

Singer, S. Fred, and Dennis T. Avery. *Unstoppable Global Warming: Every 1,500 Years*, expanded edition. New York: Rowman and Littlefield Publishers, Inc., 2008.

Solomon, Lawrence. *The Deniers*. Minneapolis, MN: Richard Vigilante Books, 2008.

Sommerhoff, G. "The Abstract Characteristics of Living Systems." In *System Thinking*, edited by F. E. Emery. Middlesex: Penguin, 1969.

Spencer, Roy W. *Climate Confusion*. New York: Encounter Books, 2008.

Stock, Gregory. *Redesigning Human Beings: Choosing Our Genes, Changing Our Future*. Boston, Houghton Mifflin, 2003.

Stockman, Lauren J., et al. "SARS: Systematic Review of Treatment Effects." 8 Oct. 2007. http://medicine.plosjournals.org/perlserv/?request=get-document&doi=10.1371journal.pm.

Stroll, Avrum. "Logical Positivism." In *The Columbia History of Western Philosophy*, edited by Richard Popkin. New York: Columbia University Press, 1999.

Studer, Raymond. "Human System Design and the Management of Change." In *General System*, vol. XVI, edited by Ludwig von Bertalanffy and Anatol Rapoport. Pocklington, York, UK: Society for General Systems Research, 1971.

Tames, Richard L., ed. *Documents of the Industrial Revolution 1750–1850*. London: Hutchinson Educational Ltd., 1971.

Tarnass, Richard. *The Passion of the Western Mind*. New York: Harmony Books, 1991.

"The Problem of Antibiotic Resistance." *NIAID Factsheet*. April 2006. http://www.niaid.nih.gov/factsheet/antimicro.htm.

"Theo Colborn Receives the Rachel Carson Leadership Award." 19 June 2008. http://www.pmac.net/nctc.htm.

Tostevin, Robert M. "Living without the Mastery of Nature: Beyond the Spell of Enlightenment." Diss. York University, 1977.

Toulmin, Stephen. *Foresight and Understanding*. New York: Harper & Row, 1961.

"Tuberculosis." *WHO Factsheet*. March 2007. http:///www.who.int/mediacentre/factsheets/fs104/en/index.html (19 Sept. 2007).

Uherek, Elmar. "Chlorofluorocarbons (CFCs) and the Ozone Hole." 20 Feb. 2006. http://www.atmosphere.mpg.de/enid/1z2.html (7 Aug. 2008).

Valentine, Vikki. "Why TB Remains a Modern and Deadly Problem." 19 Sept. 2007. http://www.npr.org/templates/story/story.php?storyId=10551019.

Vogel, Sarah. "Battles of Bisphenol A." 26 June 2008. www.defendingscience.org/case_studies_Battles_Over-Bisphenol_A.cfm.

Voltaire. "Letters on the English." In *Main Currents in Western Thought*, edited by Franklin Le Van Baumer. New Haven: Yale University Press, 1978.

Wade, Nicholas. "Couch Mouse to Mr. Mighty by Pills Alone." *New York Times*. 1 Aug. 2008. http://www.nytimes.com/2008/08/01science/01muscle.html?em.

_____. *Life Script*. New York: Simon and Schuster, 2001.

Waldrop, M. Mitchel. *Complexity: The Emerging Science at the Edge of Order and Chaos*. New York: Simon & Schuster, 1992.

Watson, James. *The Double Helix*. New York: Atheneum, 1968.

Weart, Spencer. *The Discovery of Global Warming*. Cambridge, MA: Harvard University Press, 2003.

Weinberg, Alvin. "Impact of Large-Scale Science in the United States." *Science* 134 (21 July 1961).

Westfall, Richard S. *The Construction of Modern Science: Mechanisms and Mechanics*. Cambridge: Cambridge University Press, 1971.

White, Lynn, Jr. *Machina ex Deo: Essays in the Dynamics of Western Culture*. Cambridge, MA: MIT Press, 1968.

White, Michael. *Isaac Newton: The Last Sorcerer*. New York: Basic Books, 1997.

Whitehead, Alfred North. *Science and the Modern World*. New York: The Free Press, 1967.

Wiener, Norbert. *Cybernetics: Or the Control and Communication in the Animal and the Machine*. 2nd ed. Cambridge, MA: MIT Press, 1961.

_____. *The Human Use of Human Beings: Cybernetics and Society*. Garden City, NY: Doubleday & Company Inc., 1954.

Winner, Langdon. *The Whale and the Reactor: A Search for Limits in an Age of High Technology*. Chicago: University of Chicago Press, 1986.

Wisdom, J. O. "Four Contemporary Interpretations of the Nature of Science." *Foundations of Science* 1, no. 3 (1971).

Wivel, Nelson A., and LeRoy Walters. "Germ-line Gene Modification and Disease Prevention: Some Medical and Ethical Perspectives." *Science* 262 (22 October 1993).

Yates, Frances. *The Rosicrusian Enlightenment*. London: Routledge and Kegan Paul, 1972.

Index